Complexity Management with
the K-Method

Daniel Kossmann • Donald Kossmann

Complexity Management with the K-Method

Price Structures, IT and Controlling for Procurement of Packaging Materials

 Springer

Daniel Kossmann
i-TV-T AG
Köln, Germany

Donald Kossmann
Research
Microsoft
Kirkland, Washington
USA

Translated from the German by the authors. Title of the original German Edition: Komplexitätsmanagement mit der K-Methode. (c) Springer-Verlag Berlin Heidelberg 2015

ISBN 978-3-662-48243-8   ISBN 978-3-662-48244-5   (eBook)
DOI 10.1007/978-3-662-48244-5

Library of Congress Control Number: 2015952310

Springer Heidelberg New York Dordrecht London
© Springer-Verlag Berlin Heidelberg 2016

Printed on acid-free paper

Springer-Verlag GmbH Berlin Heidelberg is part of Springer Science+Business Media (www.springer.com)

Start where you are.
Use what you have.
Do what you can.
Arthur Ashe

Preface

Packaging materials are increasingly causing problems, especially in the Fast Moving Consumer Goods industry (FMCG). This is because the variety of versions (complexity) caused by country and flavour versions has increased as well as the number of versions required for promotions and specific packaging for trade partners. This complexity is an increasing problem for the supply chain, especially for strategic procurement and material planning.

Measures such as the introduction of multilingual labels or the harmonisation of brands across countries have limited the explosion of variants but have not addressed the root cause. With the K-Method which we describe in this book, we want to introduce an entirely new way to manage the situation in such a way that the complexity is no longer perceived as such a problem. The K-Method does not challenge the number of variants of a purchasing category but shows an efficient method to handle the complexity of so many different materials. The benefits are evident when it comes to negotiating prices with the supplier, making call-offs by material planners, invoice control and reporting.

The K-Method is based on technology that has already been around for a while. However, it is rarely applied in practice to the degree explained in this book. You may find approaches you will not only be familiar with but perhaps are already using on a daily basis. With the K-Method, we provide the incentive to walk the full distance. Only then will you be able to take full advantage of the K-Method when handling your packaging materials portfolio.

The K-Method was developed in the end of the 1990s by the authors at Unilever and was tested with corrugated outers as a pilot project. Later, the K-Method was rolled out as a part of a consulting project at LR Health & Beauty Systems in Aalen, Germany. There, the categories labels and tubes were migrated to the K-Method.

The authors know that parallel to their own research, other people are also working on similar approaches, e.g., for printed foils. However, this book is the first to cover the complete approach.

We know that your time budget is tight, so we have tried to be as brief as possible. We are assuming you have a working knowledge of packaging procurement and will not repeat in this book its standard terms and procedures.

We understand that introducing the K-Method is not a simple task. But we also know, once the K-Method is introduced and established, there will be no desire by the buyer or the supplier to return to traditional procurement methods. The investment is definitely worthwhile.

We want to further develop the theory of the K-Method and its implementation in practice. We welcome any kind of feedback: corrections, suggestions on how things could be better explained ideas for further development, problems or even rejections in practice. We will also appreciate any kind of feedback on projects where you have successfully introduced the K-Method or even failed to do so. You can contact us under Daniel.Kossmann@i-TV-T.de and Donald.Kossmann@i-TV-T.de.

We hope this book will lead you to greater success!

Köln, Germany   Daniel Kossmann
Zürich, Switzerland   Donald Kossmann
Autumn 2014

Acknowledgements

The authors would like to thank the following persons who have contributed to the book:

Roland Balzer, Michel Eichwald, Thomas Kircher, Beatrix Kossmann, Horst Kossmann, Iwona Kossmann, Judith Kossmann, Alexander Kreutz, Stefan Pröls, Marieke Reinders, Uwe G. Schulte, Henk Sijbring, Peter W. Smith, Hans-Peter Schmitz, Hans-Dieter Wolf and Asmus Wolff.

Introduction

This book consists of four parts which describe different aspects of the K-Method.

The first part, "Introduction", describes the K-Method using an example. This part tries to give the reader a feeling of how the K-Method and especially its price formulas work and what kind of advantages are to be expected. This part also describes the problems which could arise when a supplier begins to use the K-Method. Such problems can be expected since both sides—the buyer and especially the supplier—possibly will be reluctant to apply the K-Method—particularly when using the K-Method for the first time. This part discusses typical objections brought forward by the parties and how to mitigate them. The purpose of the first part of the book is to convince buyers and suppliers to use the K-Method in their business relationships.

The second part of the book, "Elaboration", deals with selected problems which occur when dealing with price formulas. These problems arise when buyers and suppliers are in the process of negotiating a price formula. Hence, this part of the book is relevant once buyers and suppliers have agreed to introduce the K-Method.

The third part of the book deals with the "Implementation" of the K-Method as a business process which begins with price definition and price agreement, continues with a purchase order and finally ends with invoice control. In this part of the book, the concept of the "Format Group" is introduced which next to the price formula is another important building block of the K-Method. Specification attributes for selected buying categories are listed. These attributes are the driving force behind the prices of materials in the respective categories and, therefore, should be included in the specification system of the ERP-System. The third part of the book addresses buyers who have already agreed on a price formula with a supplier and now wish to implement the full business process efficiently.

The fourth part of the book, "Theoretical fundamentals", covers specific aspects of determining the price of a packaging material, especially the price-volume dilemma when using scaled prices. The causes and problems of inconsistent prices between different packaging materials of the same category are discussed. Also, a method of deriving the price formula of any category by just using the specifications of the materials of that category and their current prices is described. The authors find the use of the K-Method for Value Analysis to be particularly important. The fourth part of the book proves some of the assumptions made in the earlier chapters and indicates the directions in which the K-Method can be developed further in the future.

Contents

Part I

Introduction

Definitions and Typical Issues

The authors assume that the reader is familiar with the definitions of FMCG, complexity and packaging materials. We repeat these definitions to avoid misunderstandings. This chapter lists the most important topics in the field of packaging materials which the authors assume the reader is familiar with from his daily experience. These are listed to coordinate the terminology with those of the reader.

1.1 FMCG

FMCG stands for "Fast Moving Consumer Goods". These are products of daily consumption which are found in supermarkets and chemists'. These products include industrially packed foods as well as products for home and personal care. In the US the abbreviation, CPG stands for "Consumer Packaged Goods". These are all expressions which describe the same product category.

One significant feature of the FMCG is its high rate of product changes. In the beginning this was only true for branded products which were supported by advertising. Today, the products of a dealer's own brand ("DOB") and no-name discounter brands are subject to high frequency of changes in their specifications. The reasons for the changes are mainly new product launches, new variants in taste and scent, and promotions. Sometimes, product changes are made very discretely when the product is produced with a new formulation which not necessarily qualifies as an improvement. In most cases the change of a product will trigger a change of packaging. This is often only a change of the printed image on the packaging material. Even the slightest change of the formulation of a cosmetic product can lead to the change of the INCI list which is typically printed on a label.

While there is a certain consistency of the chemicals, food ingredients, flavours, and perfumes that are also used in similar products, a change of packaging materials often means that the old packaging materials still held in stock after the change

© Springer-Verlag Berlin Heidelberg 2016
D. Kossmann, D. Kossmann, *Complexity Management with the K-Method*,
DOI 10.1007/978-3-662-48244-5_1

cannot be used for other products and must therefore be destroyed. This is especially the case for printed packaging.

1.2   Complexity

The complexity of a purchasing category (in the following called "category") is defined by the number of different SKU variants ("Stock Keeping Units"). The complexity of a category increases the more often the specifications of an SKU are changed. This means that a measure of the complexity of a category is the number of SKUs multiplied by the number of specification changes in the form of new SKUs, existing SKUs petering out as well as changes of SKUs.

The management of complexity of packaging materials consists of three parts:

Specification	Specification of the packaging materials: This is an agreement between the Packaging Development Department and the supplier on the technical specifications using the specifications of the Marketing Department and the design of an agency. The specifications often trigger a mould development and print proofing.
Purchasing	Strategic buying: supplier selection and price agreement with the supplier.
Material Planning	Operative buying: Call-off of materials from the supplier referencing frame contracts while paying special attention to future specification changes.

The most intriguing elements are, of course, the specification changes of SKUs which are the greater part of added value. However, only the strategic and operational buying aspects are covered in this book. When purchasing materials the focus is only on the cost aspect of purchasing materials in the sense of purchasing prices and administration costs of the purchasing process. Hence, supplier innovations and supplier performance in respect of punctuality and complete deliveries (OTIF—on time, in full) as well as quality aspects are not covered by the K-Method. This does not mean that the authors do not rate these aspects as important, but a good "OTIF in Spec." score should be a precondition for applying the K-Method. In a sense, this is a fair assumption because over the past years, suppliers who have not proven to be reliable have been pushed out of the market by the strong competition. Furthermore, supplier innovations, meaning new kinds of packaging materials based on new production technologies, play a small role. This is not because FMCG producers do not welcome such innovations, but is attributable to the fact that most specification changes of packaging materials are based on existing technology. Having said this, we can focus on purchasing costs; i.e., the prices of materials sourced from third party suppliers.

1.3   Packaging Materials

To avoid misunderstanding, we would like to list below the most important categories of packaging materials (Table 1.1).

Though, this list is not complete, it covers the most important purchasing categories. With this book the reader will become acquainted with the main principles of the K-Method and will be able to apply the K-Method to those purchasing categories he is responsible for, even those not listed above.

Table 1.1 Purchasing categories—packaging materials

	Material	Printed	Unprinted	Technology
Bottle	PE, PP, HDPE, PVC	(X)	X	Extrusion
Bottle	PET		X	Injection/Extr.
Bottle, Jar	Glass		X	Glass
Jar	Plastic	(X)	X	Injection
Lid	Plastic	(X)	X	Injection
Lid	Plastic	X		Vacuum forming
Tray	Plastic	X	X	Vacuum forming
Closure	Plastic		X	Injection
Folding carton	Carton	X	(X)	Die-cut/Glue
Folding carton	Corrugated carton	X	(X)	Die-cut/Glue
Case	Corrugated carton	X	(X)	Die-cut/Glue
Display	Corrugated carton	X		Die-cut/Glue
Label	Plastic	X		all printing tec.
Label	Paper	X		all printing tec.
Foil	Plastic	X		all printing tec.
Tube	Plastic	X		all printing tec.
Tube	Laminate	X		all printing tec.
Net	Plastic, natural mat.		X	Weaving
Can	Aluminium, tin plate	X	X	Welding / Extr.
Pump	Plastic		X	Welding / Extr.

1.4   Major Issues

In order to tackle complexity management for packaging materials we would like to structure it into eight questions:

Feedstock Materials	How are new prices negotiated when market prices for feedstock materials have changed?
Internal Benchmarking	How do the prices of different materials with different specifications in the same purchasing category relate to each other?
Tender	How can price offers from several suppliers covering all the materials of a purchasing category be obtained?

New Prices	How can a price for a new packaging material be obtained, taking into account the volume already supplied by the supplier?
Lot Sizes	How can the lot size of a packing material to be called-off from the supplier be determined?
Combination	How can cost advantages the buyer might have with the supplier be applied efficiently or even automatically if packaging materials need to be produced by the supplier in a certain sequence?
Moulds	How are moulds dimensioned and the number of cavities determined?
Controlling	How is a meaningful reporting system for packaging materials installed?

For all these issues solutions already exist and are commonly used. However, we believe for all of these eight issues the K-Method offers a better solution than those currently applied in the FMCG industry.

The K-Method, Example of a Price Formula

Before we explain how the K-Method is able to provide a good solution for the aforementioned "Major issues", we would like to explain first what the K-Method is and how the K-Method works.

2.1 One Single Price

The easiest way to manage complexity when purchasing packaging materials is to agree to a single price (or flat price) for all the packaging materials of a purchasing category. This means the specifications and volumes of the individual packaging materials are ignored and a single price is agreed to with the supplier for all packaging materials.

Such a single price is possible from the supplier's side only by using a mixed calculation. It is obvious that bigger packaging materials require more feedstock materials than smaller items. Hence, those bigger packaging materials will have a smaller margin for the supplier when applying a single price. It is also obvious that packaging materials produced in a high volume production run will have lower production cost than those produced in a low volume production run because the machine setup costs are spread over more packaging materials when the supplier produces in a high volume production run.

All these effects can be calculated and compensated through a mixed calculation which then leads to a single price for all packaging materials. The greatest problem with a single price is that it only works if call-off volumes and specifications remain unchanged during the contract period agreed to between the supplier and the buyer. Otherwise, the single price becomes a bet between the supplier and the buyer: if the average volume of a call-off is reduced, if the dimensions of packaging materials increase, if the number of printing colours increases, if the printing surface increases, the costs for the supplier will increase in comparison with his original calculation and he will lose some, if not all, of his profit margin. In such cases, the FMCG buyer would be the winner of the bet. On the other hand, when call-off

© Springer-Verlag Berlin Heidelberg 2016
D. Kossmann, D. Kossmann, *Complexity Management with the K-Method*,
DOI 10.1007/978-3-662-48244-5_2

volumes increase or dimensions are reduced, the FMCG buyer will lose his bet and he will pay more than he should. When using single pricing, it is already enough if packaging materials of a certain specification are called-off in higher volumes, and other packaging materials in lower volumes, to trigger a mixed effect which will be unfavourable to one of the parties.

Bets tend to come with a "Betting Insurance" in the form of increased prices. This means, the supplier will provide for future unfavourable mixes by increasing the single price as not to suffer a loss when specification changes occur that will bring him below his margin. This means that the buyer will pay upfront for something he eventually might not make use of. Therefore, single pricing is to be rejected as a pricing method for packaging materials. Not only for this reason, is single pricing not commonly used in current business relations between FMCG buyers and suppliers.

2.2 Fair Price for Each Packaging Item

2.2.1 Consistent Prices for Individual Specification Attributes

From now on we assume that every packaging item carries a fair price to avoid the negative aspects of a mixed calculation. With "fair" we mean a consistent relationship between packaging items with different specifications. To phrase this is in a more general form: when a packaging item with a certain specification is extended by a feature "A" then its price will increase by X EUR/k pcs. A pricing system for a packaging category is considered to be "fair", when every time a packaging item is extended by such an attribute "A", the respective packaging item will increase its price by X EUR/k pcs—irrespective of which other features the packaging item has already specified. This means, feature "A" will have consistently added the same amount to the price of all packaging materials which carry feature "A".

The point of view that a feature will always trigger the same increase in prices consequently leads to the scenario that every feature carries a single price. Hence, the total price of a packaging item is the sum of prices of all the features that form the specifications of a packaging item. The list of features, the individual prices for these features, and the calculation method to link the prices of individual features is called a price formula. The price formula agreed to between the supplier and the FMCG customer is an important part of the K-Method.

At this point we would like to emphasize that the price formula of the K-Method is fundamentally different from what is known as "Cost Engineering". With Cost Engineering, the aim is to identify and quantify cost types and related costs of the supplier. Typically, these are depreciation, energy and labour costs. All these cost types are not features of a packaging material and they are not treated separately in a price formula. Even with feedstock materials, the only overlap between the K-Method and cost engineering, is that the K-Method pursues a different approach: While cost engineering tries to figure out what the real supplier purchasing costs for feedstock materials are, the K-Method requests a price from the supplier for feedstock materials allowing him to include logistics cost and a margin. To put it in a nutshell: cost engineering tries to make the supplier's cost structure transparent to the buyer; the K-Method tries to produce consistent prices within a purchasing category.

2.2.2   Consistent Margin of the Supplier When Quoting Prices for Specification Features

In the chapter above we determined that a specification feature is always quoted with the same price for any packaging item. Now we wish to make sure that the supplier always calculates the same profit margin for any feature. This removes the incentive for the supplier to give preference to producing packaging materials which have high margin features while being reluctant to produce those with low margins.

This requires the supplier to be a good judge of market realities. Possibly, the supplier might have a competitive advantage when producing certain specific features. This may force his competitors to price this feature relatively high. In this case, the supplier will only pass on a small portion of his cost advantage as a price reduction to his FMCG customer while keeping most of the advantage as an extra margin for himself. On the other hand if he has a competitive disadvantage versus his competitors he may have to settle with a very low margin to be able to make offers at competitive prices for the same specification feature.

When applying the K-Method at any given point of time, the supplier will disclose neither his margin nor his costs. Therefore, the quotation of specification features at a constant supplier profit margin is rather a recommendation than a requirement.

It is very important is to apply a constant profit margin when calculating machine setup prices. Machine setups are not a specification feature of a packaging material but are an important part of the purchase order of the FMCG customer. A part of the price formula of the K-Method deals with the separate quotation of machine setups. We strongly advise that the supplier include the same profit margin (calculated on an hourly basis) when effecting machine setups as when producing regular packaging materials.

2.3   Method to Create a Price Formula

2.3.1   Selection of Specification Features

When creating a price formula, the first question is which specification features of the packaging material should be included in the price formula. Some features have no relevance when it comes to production costs, others have only slight relevance and others still are the main cost drivers of the packaging material.

In the end it is up to the supplier to decide which features are to be included in the price formula as he knows his own production process best. Feedstock materials play an important role for nearly all the packaging categories. Therefore, the feedstock material used needs to be specified as well as the amount contained in the packaging item. There is no standardised method to select the other features. However, at the end of the selection and quotation process, there should be a price formula which will always deliver to the supplier a constant margin based on turnover and time, irrespective of which specifications the customer chooses to order.

This constant margin should also be achieved when the customer decides to purchase only packaging materials with large dimensions. The same applies in a scenario in which the customer chooses to call off packaging materials with very small dimension and with very small production runs. Hence, it is still best to avoid mixed calculations between different dimensions, other features or production volumes. The supplier will always have a constant margin from the amount stemming from the price formula as a function of production time.

Of course, the price formula is a simplification versus the calculation programmes the supplier normally uses. An absolute margin consistency, therefore, cannot be achieved for the price formula. Hence, a certain margin volatility is acceptable. For the following example, we will apply a volatility of 3 % versus the absolute fair price which would be obtained from the calculation programme of the supplier.

2.3.2 Example

As an example, we would like to obtain a price formula for self-adhesive labels. We will go through the process step by step. In our example the FMCG buyer buys seven different labels from a supplier with the following annual quantities in 1000 pieces (k pcs) and prices (Table 2.1).

This is a rather simple example. In reality, a buyer may be responsible for a portfolio containing several hundred even several thousand different labels. But this small portfolio in our example serves our purpose.

Typically, prices per 1000 pieces (EUR/k pcs) have developed over time. The specifications of each label were defined at a different times and the price of that label was also agreed to with the supplier at different times. Once the price of a label has been agreed to for the first time, the prices of all the labels in a portfolio are adjusted during annual negotiations, normally by a percentage which is applied as a price change across the complete portfolio.

The supplier will typically use a sophisticated programme to calculate his prices when quoting for a label for the first time. This software determines the price taking into account, production costs, assumptions about the utilisation of assets, the market situation, customer strategy and possibly also synergy effects in the sourcing of feedstock materials and production of similar labels. The FMCG buyer has no access to this supplier calculation programme or to its algorithms, or to the

Table 2.1 Example labels—original prices

	Annual vol.	Price	Price
	in k pcs	in EUR/k pcs	in EUR
Label A	4,000	14.82	59,274.10
Label B	50	31.30	1,564.80
Label C	2,500	18.31	45,775.78
Label D	1,750	6.55	11,462.53
Label E	850	79.04	67,187.42
Label F	150	46.93	7,040.08
Label G	200	32.45	6,490.50
Sum	9,500		198,795.21

Total volume: 9,500 k pcs
Total price: 198,795.21 EUR
Average price: 20.93 EUR/k pcs

assumptions fed into it by the supplier. When applying the K-Method he does not need to. Currently, the buyer only receives the results of the supplier's calculation programme which is the price for his materials (in our example labels) in accordance with the buyer's specifications.

The big difference in prices between labels having different specifications is remarkable. Nevertheless, we would like to introduce the K-Method with the rather naive approach of a "single price". This means, that we force the supplier, at least for the sake of the method, to calculate a single price for every label, irrespective of the label's specifications. In our example the weighted average price is 20.93 EUR for 1000 labels.

This single price would be okay for both parties as long as volumes and specifications do not change, despite the massive mixed costing calculation. Unfortunately, this assumption is seldom correct. Volumes and specifications change on a regular basis, which normally triggers new negotiations between the supplier and the buyer.

It is one of the targets of the K-Method to find a price mechanism which avoids the necessity of having price negotiations when volumes or specifications change. Single pricing is obviously not the winning approach as it implies mixed costing and therefore is not sustainable. The target must be to modify the single price based on its cost drivers in such a way that the cost drivers themselves receive individual prices. This will lead to a price structure which no longer has mixed costing and is, therefore, robust against future volume and specification changes.

(i) Size

The size of a label plays an important role. Big labels are normally more expensive than small labels due to the quantity of material (substrate) used in the production process. So in a first step we isolate the size effect by converting the single price to a surface price.

Table 2.2 Example labels—surface price

	Annual volume in k pcs	Typ	Height in mm	Width in mm	Surface in m2/k pcs	Price in EUR/m2	Price in EUR/k pcs	Price in EUR
Label A	4,000	white	111	73	8.103	2.75	22.28	89,133.00
Label B	50	white	84	60	5.040	2.75	13.86	693.00
Label C	2,500	transparent	98	66	6.468	2.84	18.37	45,922.80
Label D	1,750	transparent	45	45	2.025	2.84	5.75	10,064.25
Label E	850	white	112	180	20.160	2.75	55.44	47,124.00
Label F	150	white	100	80	8.000	2.75	22.00	3,300.00
Label G	200	white	85	55	4.675	2.75	12.86	2,571.25
Sum	9,500							198,808.30

Total volume: 9,500 k pcs
Total price: 198,808.30 EUR
Average price: 20.93 EUR/k pcs

The FMCG customer uses in his specifications two different substrates: "white" and "transparent". The supplier makes an offer for labels according to the selection of substrate:

Transparent: Thickness 60 my $2.84 \ EUR/m^2$
White: Thickness 85 my $2.75 \ EUR/m^2$
(1 my = 1/1000 mm)

Now we expand the price table for the labels by the dimensions of each label in order to calculate the surface of each label. The surface of each label is multiplied by the surface price of the substrate to obtain the price of each label (Table 2.2).

In this way we obtain the same grand total and the same average price of 20.93 EUR/k pcs. We have now also found a formula to reflect future volumes and specification changes without the necessity of carrying out additional price negotiations.

When we now compare the surface prices (third column), which takes into account the different substrates and label sizes with the original prices (first column) we see big differences (last column). This suggests that from a supplier's point of view, mixed costing is still being used, almost in the same way as when applying a single price. Especially when looking at Label A (original price: 14.82 EUR/k pcs, new price: 22.28 EUR/k pcs) the buyer will feel invited to purchase this label from another supplier. Just by going to another supplier for this label, the buyer could generate savings for himself. On the other hand prices for Label G (original price: 32.45 EUR/k pcs, new: 12.86 EUR/k pcs) is very favourable. The supplier may be selling this label at a loss, so he will not be interested in selling any more labels with similar specifications.

Summary: Defining the price of a label using its surface dimensions is a better method than using single pricing. However, it is still not robust enough to be used as the

Table 2.3 Example labels—price comparison between the original price and surface prices

	Original price in EUR/k pcs	Single price in EUR/k pcs	Surface price in EUR/k pcs	Price difference in %
Label A	14.82	20.93	22.28	50.4%
Label B	31.30	20.93	13.86	-55.7%
Label C	18.31	20.93	18.37	0.3%
Label D	6.55	20.93	5.75	-12.2%
Label E	79.04	20.93	55.44	-29.9%
Label F	46.93	20.93	22.00	-53.1%
Label G	32.45	20.93	12.86	-60.4%

only price mechanism because the supplier will still be forced to use mixed costing. In the future, when volumes and specifications are changed, this will lead to new price negotiations. Thus, additional specification features need to be part of the price formula.

(ii) Machine Setup ("Setup")

Apart from the surface and due to the quantity of required feedstock materials (substrate), the machine setup and its cost play a significant role when producing labels. Typically, a price per label is agreed to between the buyer and the supplier. When the supplier sets up the printing machine, there is no revenue to balance the involved cost because during machine setup only waste instead of labels is produced. The supplier will distribute the cost of machine setup over the labels of the following production run, treating this as a kind of amortization.

As a consequence of amortizing machine setup costs, labels produced in a long print runs will be cheaper than the same labels printed in a short run. The supplier normally passes on these savings to the buyer. E.g. he offers scaled prices where he defines quantity thresholds with respectively lower prices. The more labels the buyer orders, the lower the price of each individual label will be.

Predefined thresholds are the biggest weakness of scaled pricing. A customer's modern production planning will use MRP runs to determine material requirements. Those quantities, calculated for each production run, are rarely the ones agreed to as thresholds in scaled pricing. This may lead to the obviously unfavourable, but economically optimal situation, that it is better to order a higher than needed quantity, which may not be used and will have to be disposed of, to reach the next price threshold, to minimize the total cost of the call-off. Otherwise, ordering slightly fewer quantities than required for the next threshold may increase the total purchasing costs by as much as 10 %.

In respect of machine setups, the K-Method has a different approach. Here, the focus is not on the supplier's costs which must somehow be covered, but on a single price. The fundamental idea behind the K-Method is that every machine/hour needs to generate a margin for the supplier to cover overheads and profit. It is not relevant for the K-Method if the machine is running during the time needed to produce a label

("Run") or if the machine is just in the process of preparation of production ("Setup"). In both cases the machine needs to generate a margin per hour. Again, the supplier is asked to keep his calculation clean and not to allow any form of mixed costing. This leads to the obscure observation that the supplier's margin is the same whether he is printing labels for 100 h or he is permanently setting up his printing machine during those 100 h. This could mean that the supplier was setting up his machine 50 times for 50 different label formats without printing a single label.

This requirement of having equal margins for setups as for actual print runs on an hourly basis may seem reasonable for outsiders but for suppliers this is rather weird. In any case, this requirement is a core element of the K-Method. This right of compensation for machine setups gives the customer total flexibility with respect to call-off volumes. He may use this flexibility to his own advantage rather than to the advantage of the supplier. Further on this in later chapters.

After assuming that the supplier will have a fixed margin per machine hour, the supplier will now be interested in selling as many machine hours as possible. When looking at a portfolio with a certain annual volume, he will be interested in small production runs. In such a case more machine setups are required and therefore the number of machine hours is maximized. The buyer needs to discourage the supplier from carrying out large production runs and saving machine setups while continuing to invoice machine setups which were never actually needed. This supplier's dilemma will also be discussed more thoroughly in a later chapter.

To continue with our example: the supplier will now also explicitly charge for every machine setup. For our example, we have chosen specifications requiring a wide range of printing technologies and we also selected a supplier who is able to serve these printing requirements: offset printing, flexo printing, and silkscreen printing.

At this point we need not elaborate on the different printing technologies. As an experienced buyer we accept them as part of the specifications which are provided by the Packaging Development Department. However, it is important to note that the K-Method can calculate the label cost for all kinds of printing technologies using the same print impression. In this way, Marketing can decide if the higher print quality achieved by a more expensive print technology is worth its cost on a specific packaging item. Thus, the K-Method also facilitates the Value Analysis of a packaging item. This aspect of the K-Method will be discussed in greater depth in a later chapter.

The supplier makes the following price offer for machine setups:

Offset	500 EUR + 150 m material
Flexo	300 EUR + 300 m material
Silkscreen	300 EUR + 150 m material

This means that a machine setup has a time/cost component and a material component. When setting up a machine, material is used to fine-tune the printing image. This material is wasted and must be disposed of. Typically, the waste material is expressed in metres and not square metres as it is a function of the speed of the printing machine and the time required to achieve a satisfactory printing image. The effective surface is finally calculated by multiplying the metres with the effective printing width of the printing machine which is a function of the height of the label and the number of labels printed at the same time.

Since the substrate costs are volatile and differ according to the specifications of the substrate the supplier offers the substrate separately:

| Transparent: | 60 my | 0.73 EUR/m^2 |
| White: | 85 my | 0.71 EUR/m^2 |

For the purpose of this example, the waste generated during the machine setup will be charged at this price.

At this point we would like to discuss the possibility of producing several labels at the same time ("Repeat factor"). Every time a production machine allows the production of several packaging items at the same time, this question arises. Often moulds determine the number of packaging items produced simultaneously, e.g., when using injection moulding, the number of cavities of the mould determines the output during each production cycle. When printing labels the height of a label determines the maximum number of labels produced at a time. We will assume a maximum printing width of 350 mm which will lead to the following repeat factor (Table 2.4).

This means several labels can be produced at the same time. They are placed parallel to each other. At the end of the production run, the print coil is cut into several coils in such a way that only one label placed vertically is left on each coil. Only coils with that format can be used with labelling machines in the production hall of the FMCG customer.

The coils used for printing are assembled from much bigger coils. A coil with the effective printing width can thus be used and the waste during machine setup only needs to be calculated on the basis of the effective printing width (Table 2.5).

The price of the machine setup in our example is only 9 % of the total label price. As experience has shown, this can be much higher depending on the production run and the purchasing category.

To simplify the price formula for machine setups, a certain number of printing colours can be included in the price from the start. In our example the supplier and

Table 2.4 Example labels—calculation of number of copies printed at a time

	Printing width	Label height	Repeat Factor	Effective Printing width
	max in mm	in mm	in pcs	in mm
Label A	350	111	3	333
Label B	350	84	4	336
Label C	350	98	3	294
Label D	350	45	7	315
Label E	350	112	3	336
Label F	350	100	3	300
Label G	350	85	4	340

Table 2.5 Example labels—machine setup

	Print	Print runs/ Call-off	Setup	Effective Print width	Typ	Substrate	Waste		Machine Setup			Total year
		per year	in EUR/Setup	in mm		in EUR/m2	in m	in EUR/Setup	in EUR/Setup	in EUR/k pcs		in EUR
Label A	Offset	12	500.00	333	white	0.71	150	35.46	535.46	1.61		6,425.57
Label B	Offset	1	500.00	336	white	0.71	150	35.78	535.78	10.72		535.78
Label C	Flexo	12	300.00	294	transparent	0.73	300	64.39	364.39	1.75		4,372.63
Label D	Flexo	6	300.00	315	transparent	0.73	300	68.99	368.99	1.27		2,213.91
Label E	Silkscreen	4	300.00	336	white	0.71	150	, 35.78	335.78	1.58		1,343.14
Label F	Silkscreen	4	300.00	300	white	0.71	150	31.95	331.95	8.85		1,327.80
Label G	Offset/Silks.	2	800.00	340	white	0.71	300	72.42	872.42	8.72		1,744.84
Sum		41										17,963.68

Total number of setups per year   41

Total price of machine setups   17,963.68 EUR

the buyer have agreed on three colours. For any further colour, the buyer will have to pay a supplement:

Offset   60 Euro + 70 m for each additional colour
Flexo   45 Euro + 120 m for each additional colour
Silkscreen   125 Euro + 50 m for each additional colour

This means that the number of colours determines the price of the label since additional machine setups may be required. There is a maximum number of colours which a printing machine can apply at a time. If the customer specifies more colours than this maximum, the labels need to be printed twice. This adds significant complexity for the machine setup as well as for the production process of the label. Since such specifications do not deliver additional insights into the K-Method, we have not included such an example (Table 2.6).

Table 2.6 Example labels—setup for additional colours

	Print	Colours	Setup	Extra			Width	Substrate	Waste		Setup for additional colours		Total year
		(3 col. free)	in EUR/colour	in EUR/colour	in EUR	in m/colour	in mm	in EUR/m2	. in m2	in EUR/Setup	in EUR/Setup	in EUR/k pcs	in EUR
Label A	Offset	6	60.00	0.00	180.00	70.00	333	0.71	69.93	49.65	229.65	0.69	2,755.80
Label B	Offset	5	60.00	120.00	240.00	70.00	336	0.71	47.04	33.40	273.40	5.47	273.40
Label C	Flexo	4	45.00	0.00	45.00	120.00	294	0.73	35.28	25.75	70.75	0.34	849.05
Label D	Flexo	4	45.00	0.00	45.00	120.00	315	0.73	37.8	27.59	72.59	0.25	435.56
Label E	Silkscreen	4	125.00	0.00	125.00	50.00	336	0.71	16.8	11.93	136.93	0.64	547.71
Label F	Silkscreen	4	125.00	0.00	125.00	50.00	300	0.71	15	10.65	135.65	3.62	542.60
Label G	Offset/Silks.	6/3	60.00	0.00	180.00	70.00	340	0.71	71.4	50.69	230.69	2.31	461.39
Sum													5,865.52

Total number of setups per year   41   (see table 2.5.)

Total price Setup (setup + material)   17,963.68 EUR   (see table 2.5.)

Total price Setup (additional colours)   5,865.52 EUR

In our example the three additional colours for Label G need to be paid extra. The three colours used for silk screen printing are already included in the machine setup price.

Please note that there is a special, additional setup charge for Label B which is listed with 120 EUR in the column "Extra". The reason for this additional setup charge will be explained at a later stage of this example. For the time being just remember that for Label B there are additional costs for the machine setup which are charged above and beyond the five colours for the offset printing.

After having determined the setup charges, we would like to discuss the almost philosophical but definitely technical question of whether all the activities of a machine setup need to be carried out every time a different label is printed. The experienced label buyer will know that there may be several variants of each label. When switching production from one variant to another, fewer setup activities are required at the printing machine; for example, when a label has several different language variants and the label has been designed so cleverly that when changing the language only one colour is affected. In this case, only one printing plate needs to be changed and all the other colours and printing plates can stay as they are. Such a plate change, known as "changeover", will require a separate price. It will save the buyer money if he orders labels in such a way that many setups can be converted to changeovers, as changeovers are cheaper than full setups.

In our example, all the labels are printed in several different variants. To simplify the example, all the variants are printed at each production run. Thus if a label has four variants, one setup must be paid for and there are three changeovers (Table 2.7).

Table 2.7   Example labels—setup and changeover

	Print	Variants	Changeover	Changeover for additional variants		Total year
			in EUR/colour	in EUR/Setup	in EUR/k pcs	in EUR
Label A	Offset	4	60.00	180.00	0.54	2,160.00
Label B	Offset	4	60.00	180.00	3.60	180.00
Label C	Flexo	4	50.00	150.00	0.72	1,800.00
Label D	Flexo	4	50.00	150.00	0.51	900.00
Label E	Silkscreen	10	150.00	1,350.00	6.35	5,400.00
Label F	Silkscreen	2	150.00	150.00	4.00	600.00
Label G	Offset/Silks.	1	210.00	0.00	0.00	0.00
Sum						11,040.00

Total number of setups per year	41	(see table 2.5)
Total price Setup (setup + material)	17,963.68 EUR	(see table 2.5)
Total price Setup (additional colours)	5,865.52 EUR	(see table 2.6)
Total price Changeover	11,040.00 EUR	
Total price Setup & Changeover	**34,869.19 EUR**	

This means that a high number of label variants can cause the buyer high costs for changeovers. But they are still lower than if a full machine setup is charged. To avoid these high changeover costs, many FMCG companies try to place as many languages as possible on a single label ("language clusters"). Such language clusters not only avoid changeover charges but also increase the flexibility in logistic terms by increasing the flexibility in stock management. If, on short notice, the volume for one country is increased while the volume of another country is decreased, this will not affect the requirement for labels as such multi-language

Table 2.8 Example labels—setup and surface price for printing

	Volume	Call-offs	Volume	Setup & Changeover	Print surface price	Surface	Price	
	in k pcs	per year	in k pcs/cll-off	in EUR/call-off	in EUR/m2	inm2/k pcs	in EUR/call-off	in EUR/k pcs
Label A	4,000	12	333.3	945.11	2.27	8.10	7,070.10	21.21
Label B	50	1	50.0	989.18	2.27	5.04	1,560.64	31.21
Label C	2,500	12	208.3	585.14	2.34	6.47	3,740.84	17.96
Label D	1,750	6	291.7	591.58	2.34	2.03	1,974.76	6.77
Label E	850	4	212.5	1,822.71	2.27	20.16	11,537.43	54.29
Label F	150	4	37.5	617.60	2.27	8.00	1,297.90	34.61
Label G	200	2	100.0	1,103.11	2.27	4.68	2,163.25	21.63

labels can be used for both countries. In a later chapter we will discuss the matter of whether multi-language labels are really such a good idea.

After putting a price tag to machine setups and changeovers, we need a new price for the actual printing of the surface which now excludes machine setups and changeovers. Our sample supplier quotes the following prices:

| Transparent: | 60 my | 2.34 EUR/m^2 |
| White: | 85 my | 2.27 EUR/m^2 |

The new version of the price formula is the following (Table 2.8).

Table 2.9 Example labels—price comparison: original price, surface price, setup and surface price for printing

	Original price	Surface price	Setup + surface price	Price difference	Volume	Setup + surface price
	in EUR/k pcs	in EUR/k pcs	in EUR/k pcs	in %	in k pcs	in EUR
Label A	14.82	22.28	21.21	43.1%	4,000	84,841.25
Label B	31.30	13.86	31.21	-0.3%	50	1,560.64
Label C	18.31	18.37	17.96	-1.9%	2,500	44,890.04
Label D	6.55	5.75	6.77	3.4%	1,750	11,848.54
Label E	79.04	55.44	54.29	-31.3%	850	46,149.72
Label F	46.93	22.00	34.61	-26.3%	150	5,191.61
Label G	32.45	12.86	21.63	-33.3%	200	4,326.50
Sum						198,808.30

| Total | 198,808.30 EUR |
| Average | 20.93 EUR/k pcs |

The price development, after another step of the K-Method has been added, can be seen in Table 2.9.

By introducing Setup prices, the total price for the annual label volume has not changed. The average price across all labels remains at 20.93 EUR/k pcs. However, this price now follows a certain pricing strategy and should be differentiated. Unfortunately, single prices for labels are still quite far away from the original prices. This is probably because we have not achieved the original target of providing a pricing system without mixed costing. To reach our target we have to differentiate the actual printing of the labels better than we have done up to now by using just the surface.

(iii) Printing ("Run")

We already isolated the price for the substrate when we calculated the waste during machine setup. For the actual printing we should apply the same price for the substrate and ask for a separate quotation for the actual print run:

Offset	0.25 EUR/m
Flexo	0.49 EUR/m
Silkscreen	0.90 EUR/m
Offset/Silkscreen	1.25 EUR/m

Most printing machines run at a constant speed. Time—since we want to generate a constant profit margin per machine hour—can, therefore, be a function of metres produced by the machine. The supplier thus quotes the print run with 1 m as the base measurement unit.

He will also charge for an extra service called hot foil stamping. This hot foil stamping delivers shiny silver or gold effects on the labels.

Hot foil stamping 0.30 EUR m

As you might remember, Label B was charged with an extra 120 EUR for its machine setup. This charge was for hot foil stamping (Table 2.10).

Table 2.10 Example labels—calculation of run prices for printing

	Annual volume *in k pcs*	No of call-offs	Call-off volume *in k pcs*	No of labels *in pcs*	Label width *in mm*	Production length *in m*	Price printing *in EUR/m*	Special	Price special *in EUR/m*	Total printing *in EUR/k pcs*
Label A	4,000	12	333	3	73	8,111.1	0.25	-	-	6.08
Label B	50	1	50	4	60	750.0	0.25	Hot foil stamping	0.30	8.25
Label C	2,500	12	208	3	66	4,583.3	0.49	-	-	10.78
Label D	1,750	6	292	7	45	1,875.0	0.49	-	-	3.15
Label E	850	4	213	3	180	12,750.0	0.90	-	-	54.00
Label F	150	4	38	3	80	1,000.0	0.90	-	-	24.00
Label G	200	2	100	4	55	1,375.0	1.25	-	-	17.19

Now we have reflected the printing costs in an adequate price which was mainly determined by the number of labels printed at the same time, in parallel, and the printing technology used.

The only part which is missing is the substrate on which the labels are printed. We already mentioned the supplier's quotation for this feedstock material (Table 2.11).

The printing price and feedstock materials are now included. Both are dependent on the call-off volume also known as the Run price.

Table 2.11 Example labels—calculation of the run price including printing and feedstock materials

Label height	Label width	Call-off volume		Typ	Substrate	Substrate	Printing	"Run"	Total year
in mm	in mm	in k pcs	in m2		in EUR/m2	in EUR/k pcs	in EUR/k pcs	in EUR/k pcs	in EUR
Label A	111	73	333	8.10	white	0.71	5.75	6.08	11.84	47,345.85
Label B	84	60	50	5.04	white	0.71	3.58	8.25	11.83	591.42
Label C	98	66	208	6.47	transparent	0.73	4.72	10.78	15.50	38,754.10
Label D	45	45	292	2.03	transparent	0.73	1.48	3.15	4.63	8,099.44
Label E	112	180	213	20.16	white	0.71	14.31	54.00	68.31	58,066.56
Label F	100	80	38	8.00	white	0.71	5.68	24.00	29.68	4,452.00
Label G	85	55	100	4.68	white	0.71	3.32	17.19	20.51	4,101.35

Table 2.12 Example labels—setup, changeover and run

	Setup	Change-over	Run	Setup, CO, & Run	Original price	Surface price	Setup + surface price	Setup, CO & Run	price difference
	in EUR/k pcs	in EUR/k pcs	in EUR/k pcs	in EUR/k pcs	in EUR/k pcs	in EUR/k pcs	in EUR/k pcs	in EUR/k pcs	in %
Label A	2.30	0.54	11.84	14.67	14.82	22.28	21.21	14.67	-1.0%
Label B	16.18	3.60	11.83	31.61	31.30	13.86	31.21	31.61	1.0%
Label C	2.09	0.72	15.50	18.31	18.31	18.37	17.96	18.31	0.0%
Label D	1.51	0.51	4.63	6.66	6.55	5.75	6.77	6.66	1.6%
Label E	2.22	6.35	68.31	76.89	79.04	55.44	54.29	76.89	-2.7%
Label F	12.47	4.00	29.68	46.15	46.93	22.00	34.61	46.15	-1.7%
Label G	11.03	0.00	20.51	31.54	32.45	12.86	21.63	31.54	-2.8%

In the context of the volume scenario of our example, when we merge Setup prices, Changeover prices, Run prices we obtain the following prices for each label (Table 2.12).

When we compare these new prices with the original prices, we see that the new prices have a maximum deviation of 2.8 %. Thus, they are all within the 3 % tolerance we were aiming at to avoid mixed costing at the supplier's end. However, in this example and generally speaking, we do not know the supplier's margins. We have assumed that the supplier originally applied the same profit margin (in %) for each label when he originally quoted prices for the labels. Consequently, it cannot be assumed that the new prices which vary a maximum of 3 % from the old quotations guarantee a maximum deviation of 3 % of the margin, but the real deviation of the profit margin will be close to these 3 %.

2.3.3   Summary

From experience we know that the change from conventional individual pricing to a price formula will not yield such a harmonic picture as shown in our example. There will be spikes which enrich negotiations between supplier and buyer. Spikes towards higher original prices are probably labels which have not enjoyed the

Table 2.13 Example labels—price formula with prices for specification features

			Offset	Flexo	Silkscreen BuM
	Maximum print width		350	350	350 mm
Setup	*3 colours*	Setup	500.00	300.00	300.00 EUR
		Material	150.0	300.0	150.0 meter
	each additional	Setup	60.00	45.00	125.00 EUR
	color	Material	70.0	120.0	50.0 meter
		Hot foil	120.00	-	- EUR
Changeover	*each colour*	Setup	60.00	50.00	150.00 EUR
Run		Printing	0.25	0.49	0.90 EUR/meter
		Hot foil	0.16	-	- EUR/meter
		Substrate white			0.71 EUR/m2
		transparent			0.73 EUR/m2

increased attention of the buyer, have not been the focus of negotiations and are, therefore, prices that have been carried forward over time. The new lower price, stemming from the K-Method, is a more than welcome saving for the buyer. The spikes at the lower end of the price scale, where originally quoted labels proved to be cheaper than now with the K-Method, are much more interesting. What enabled the supplier to quote such low prices to begin with? Very often, labels with the highest annual volumes are those which spike toward the bottom. Those labels are very often considered to be "strategic". This means that the supplier offers more competitive prices for these items as for those items used by the buyer to benchmark suppliers. Very often, the supplier accepts razorblade thin margins for these strategic items to recover his margin with higher prices for the other labels. In this case the buyer has a realistic chance to calibrate the entire price formula of the K-Method using those strategic labels as reference points. This will bring down the whole price level for all the labels and yield savings for the buyer.

In short: by determining prices for each specification feature, the K-Method was able to define a price frame. For our example the price frame was as shown in Table 2.13.

With this price frame or price formula all the prices for labels can be (newly) calculated (again). This also applies to future labels of which the buyer does not yet know the specifications.

For our example of seven different labels with 29 different variants, the effort to create such a price formula may seem to be overdone. But in practice, the portfolios may have several hundred or several thousand different label specifications and variants. And for these portfolios the K-Method with its price formulas is a very efficient method to negotiate, determine and manage prices, even when in practice the price formula will have twice as many parameters as in our example.

Solutions for Major Issues Using the K-Method

3

In this chapter we would like to return to the issues already discussed in Sect. 1.4. We would like to demonstrate how these problems can be tackled using the K-Method.

3.1 Feedstock Materials

As the first of eight issues, we would like to discuss feedstock materials and how they should be integrated in the packaging prices.

Producers of packaging materials use raw materials to produce packaging materials. These materials are known as feedstock materials. Feedstock materials are typically commodities such as paper, plastic polymers or aluminium. Most of these commodities are not traded at commodity stock exchanges. Thus not all players have the same market information. Nevertheless, market activities are well known by all suppliers of packaging materials. There are industrial federations (e.g., VWD for paper) and specialized market research companies (e.g., Platts for polymers) which regularly conduct market surveys by polling transaction prices. Those published market volumes and market prices are considered to be reliable market information. Both, FMCG buyers as well as suppliers observe these markets very closely. They know that they cannot influence those markets because of their sheer size, but they also know that their correct prediction will have great influence on the bottom line of their companies.

Despite the fact, that in the very long run these market prices for feedstock materials constantly increase as a result of inflation, the markets are subject to small multiannual periods of significant volatility. These feedstock markets, therefore, can be co considered cyclical markets. Considering that feedstock materials represent 20–60 % of costs for the supplier, it is clear that the final price of a packaging material has a high correlation with the purchase prices of feedstock materials.

© Springer-Verlag Berlin Heidelberg 2016
D. Kossmann, D. Kossmann, *Complexity Management with the K-Method*,
DOI 10.1007/978-3-662-48244-5_3

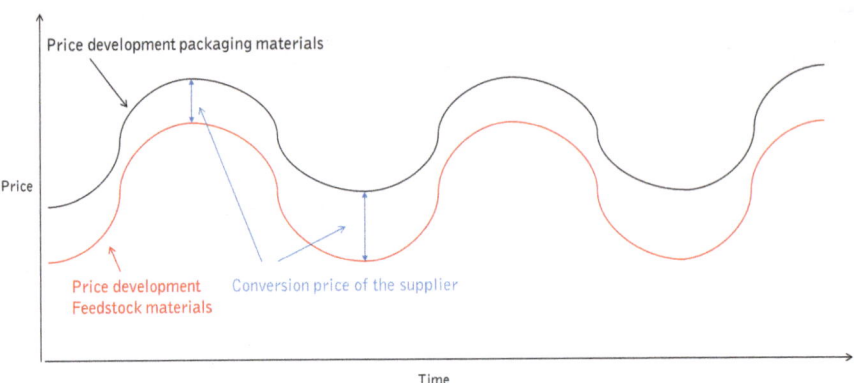

Chart 3.1 Price cycles in dependence of feedstock materials

Chart 3.1 shows the price development of a typical purchasing category. The price of feedstock materials is the major driver for the price development of packaging materials, but this is not necessarily always the case.

In an ideal market place, the supplier would pass on price changes of feedstock markets 1:1 to his FMCG customer. This would mean that the price volatility of feedstock markets would have no economic impact for him. He would thus protect his profit margin. As a consequence, in respect of feedstock materials, the supplier would charge cost prices for packaging materials to the FMCG customer rather than market prices. This idea market place would only function if the FMCG customer, too, can pass on 1:1 the price changes of packaging materials to the trade and eventually to the consumer. This would mean that the consumer would have to bear the cycles of feedstock materials—for good and for bad. Of course, this is an ideal situation which cannot be taken into account because in the end, shelf prices in supermarkets are driven by market prices rather than cost prices. Neither the FMCG producer nor the supermarket wants to pass on to the consumer savings stemming from lower sourcing costs. In the case of a supermarket deciding to lower prices, it will be part of a strategic marketing campaign and only incidentally triggered by lower feedstock prices. The FMCG producer generally does not want to make any price concessions to the retail trade unless it is a result of tough negotiations.

Eventually, this leads to the situation that price changes of feedstock materials—irrespective of direction—trigger first negotiations between the supplier and the FMCG buyer and then between the FMCG sales person and his trade partner. This is obviously not very efficient.

These negotiations lead to the conclusion that, in fact, the FMCG producers purchase materials at the lowest price when the feedstock materials are at their highest price, and vice versa, the FMCG producer purchase materials at the highest price when the prices for feedstock materials are the lowest. This seems to be paradox but it is better understood when seeing the matter from another perspective: If the buyer of a packaging material only buys the service of converting a feedstock material into a packaging material and pays for the feedstock material separately, then he will be paying the highest price for this service when the price of the feedstock material is the lowest. This situation arises because during negotiations between the supplier and the FMCG buyer, the supplier will never be able to fully recuperate the cost increases he is exposed to if the price of feedstock materials increases. His margin will thus decrease. On the other hand, when prices for feedstock materials decrease again, the FMCG buyer will not be able to persuade the supplier to pass on all the benefits to him. Some benefits will remain with the supplier and will improve his margin.

In the long term, this negotiation routing does not create value for any of the parties and requires a substantial amount of resources on both the supplier and buyer sides. Of course, this problem has been identified over time by both parties. It has been mitigated by various methods from simple negotiations ("arm length") to applying feedstock index clauses as part of the frame contracts which automatically trigger negotiations, once feedstock materials pass a certain threshold. In some cases agreements carry a calculation method how the change of feedstock prices automatically translates into new prices for packaging materials.

3.1.1   The K-Method for Feedstock Materials

The K-Method recognizes feedstock materials as a substantial element of the price of a packaging material. Contrary to Best Proven Practice (BPP), the K-Method does not impose automatic price changes. Also, reference to a publicly available price index of feedstock materials is not necessarily required. Instead, the supplier offers the feedstock material to the buyer as a separate commodity which the buyer purchases from the supplier. This price for the feedstock materials cannot be the price which is publicly available from industrial federations or market research companies since the supplier must also take into account logistic costs and overheads as well as a profit margin. The supplier will, therefore, offer the feedstock material with a surcharge. We are no longer talking about costs to be carried by the supplier, but about a price offered to the FMCG buyer. Of course, the FMCG buyer will know the average market price for the feedstock materials, but he will never discover at what price the supplier really purchases his feedstock materials. Perhaps the supplier enjoys the benefit of vertical integration, meaning that he is also the producer of feedstock materials. In this a case, the supplier will buy his feedstock materials well below market prices. Perhaps the supplier has agreed to long term prices with his supplier and is, therefore, not exposed to market price volatility during this period. Since the FMCG buyer does not have access to the actual price the supplier is paying for feedstock materials, he will also not know how much the supplier is charging him for logistics, overheads, and profit. Hence, the FMCG buyer is now purchasing feedstock materials at a new, very special market price and the supplier becomes in a sense a trader of feedstock materials.

The supplier must calculate the price for feedstock materials very carefully. It is a principle of the K-Method that the feedstock trading option is just a module. This means, that the FMCG buyer will always consider buying feedstock materials directly from the feedstock materials market and will provide these feedstock materials to the supplier asking him to convert them into packaging materials. Such an approach requires quite some effort from the buyer. In many cases the buyer does not have the same purchasing volume for his feedstock materials as the supplier who can combine this volume with the volumes of other FMCG customers. If the buyer really wants to buy feedstock materials himself, he may end up with a higher price just for the feedstock materials than the feedstock materials offered by the supplier which already includes his mark-ups. However, it is not the intention of the supplier to calculate his mark-ups for feedstock materials in such a way that it is not worthwhile for the buyer to source these materials himself. The true competitor in the price calculation for feedstock materials is not the buyer but all the other suppliers of packaging materials the buyer can choose to order from.

An important part of the K-Method is that it requires the supplier of packaging materials to make an explicit quotation for feedstock materials as a part of the total pricing system. This means every feedstock material has its own price and is offered in the same base unit of measurement ("BuM") as the supplier himself is offered by his feedstock supplier, e.g., EUR/kg for polymers. Hence, the BuM of the feedstock is not the same BuM of the packaging material sold to the buyer which is normally EUR/k pcs for, say, plastic bottles.

This also means that treating feedstock materials as a lump sum over a packaging category portfolio is not good enough. This will lead again to mixed costing since individual packaging items have different amounts of feedstock materials in volume and in percentage of the packaging price. Eventually, iterative mistakes are generated and this will invalidate the whole approach (Table 3.1).

Looking at our original example, the price of the feedstock materials used is in the range of 10.5 and 39.2 %. The average is 20.0 %, but the weighted average is 26.1 %. In the In the case of our example, mixed costing would have a strong influence if the average value of feedstock materials can be assumed across the entire range of labels. Adding to the problem is the fact that, normally, it is not the

Table 3.1 Example labels—feedstock materials as a percentage of total packaging item cost

	Annual volume	Price original	Price formula	of which feedstocks		Price total	Feedstocks
	k pcs	in EUR/k pcs	in EUR/k pcs	in EUR/k pcs	in %	in EUR	in EUR
Label A	4,000	14.82	14.67	5.75	39.2%	58,687.23	23,012.52
Label B	50	31.30	31.61	3.58	11.3%	1,580.60	178.92
Label C	2,500	18.31	18.31	4.72	25.8%	45,775.78	11,804.10
Label D	1,750	6.55	6.66	1.48	22.2%	11,648.91	2,586.94
Label E	850	79.04	76.89	14.31	18.6%	65,357.41	12,166.56
Label F	150	46.93	46.15	5.68	12.3%	6,922.40	852.00
Label G	200	32.45	31.54	3.32	10.5%	6,307.58	663.85
Sum	9,500					196,279.92	51,264.89

Total price   196,279.92 EUR
- of which feedstock materials   51,264.89 EUR
Average amount for feedstock materials   26.1%

customer who determines the average according to his portfolio, but the supplier who provides the average using all the labels he produces for all of his customers. At this point the K-Method has a much higher accuracy since it determines the amount of feedstock materials for each single packaging item, in our example: labels. To boil it down to a rule of thumb: the more complicated a packaging item is designed, the more the buyer will be at a disadvantage when negotiating average feedstock percentages across his portfolio. This applies when feedstock materials increase in price because the supplier will typically overstate the amount of feedstock materials used.

In any case, it is disturbing but mathematically correct that the percentage of feedstock materials used in a packaging material changes whenever prices for feedstock materials change. If e.g., the price of a feedstock material increases by 11.0 % and the cost block for the feedstock materials of that packaging item is 40.0 % according to the supplier, it would justify a request by the supplier to effect a price increase of 4.4 % ($=40.0 \times 11.0$ %) for that material. However, all the parties involved should be aware of the fact that once this request by the supplier is accepted, the percentage for feedstock materials will increase from 40.0 to 42.5 %. During subsequent negotiations, the supplier will probably again make his argument on the basis of a 40 % share of feedstock materials thus giving the buyer a forward-running advantage. The supplier sees this fully compensated not only when prices fall again, but also in general because he provides for too large a share of feedstock materials.

The bottom line remains unchanged: negotiations in the context of feedstock materials are arduous without delivering productive results. The buyer has to make considerable efforts to improve his market intelligence in order to conduct these difficult negotiations. The pure negotiation of feedstock materials without looking at percentages is negotiation at the core of the problem. Such negotiations are much easier to conduct for both sides and are, therefore, more efficient.

3.2   Internal Benchmarking

A fundamental question for every FMCG buyer is: "How do purchasing prices compare to those of competitors?" This question will normally never be answered because the prices at which competitors purchase their own materials are not known and are not disclosed by their suppliers. Since this question cannot be answered externally, benchmarking is not possible. Anyhow, it would not be very helpful to look at competitors' prices of packaging materials because every FMCG producer buys packaging materials with different specifications, different volumes, different dimensions and different features. Thus, with the current setup, external benchmarking is not possible, and if it is done anyway, it will require many assumptions yielding imprecise results.

Internal benchmarking is, however, much more interesting. How do the prices of packaging materials in a purchasing category relate to each other? This question is especially interesting when different packaging materials are supplied by the same supplier.

Since packaging materials have different production volumes and different specifications, it is no simple task to compare prices. What is the price impact of different surfaces when ordering folding cartons? What about an additional silkscreen print for a laminated tube? What is the impact of adding hot foil stamping on a label? These questions normally cannot be answered since the individual features of a packaging item do not carry individual prices. The negotiating parties usually have a gut feeling as to how prices change when the specifications change. But they do not have explicit price rules for this. To obtain the price difference stemming from different volumes, scaled pricing is often agreed to. Sometimes, order fixed prices ("Setup") and production run cost ("Run") are agreed to.

3.2.1   The K-Method for Internal Benchmarking

For the internal benchmarking of the portfolio of a supplier, it is best to use a price formula. Of course, the ideal situation would be to take a price formula which has been agreed to with the supplier. However, in this case, all the supplier's prices are already consistent to each other because of the price formula. Hence, spikes cannot be determined because none can appear.

In most cases no price formula will be available. In this case, it is possible to use the price formula of another supplier who delivers another portfolio of the same category or another region. With such a price formula, the spikes can be determined as well as the general price level of the supplier because it is possible to compare the actual total annual purchasing costs of the portfolio using actual prices with the price formula prices.

If it is not possible to obtain a supplier price formula, a price formula can be derived from the current prices of the different packaging materials of a category and its corresponding specifications. This derivation will use a set of linear equations which are explained in a later chapter. It is important to know that these linear equations are normally not solvable since, usually, it is not possible to obtain consistent pricing of specifications features across a category. However, a sufficient approximation can be calculated and thus show the spikes of the current prices settings.

The use of a reference price formula is much easier to use than the calculation of a price formula using linear equations. The reference price formula can be adjusted in such a way that the total annual costs of the portfolio are identical to the effective annual costs using current individual packaging prices. A collection of reference price formulas is not part of this book but available from the authors.

Once a price formula has been adjusted to a portfolio, by arithmetic interpolation or because it is the price formula which was agreed to between the supplier and the buyer, it can be used as a benchmarking reference. The portfolio which has been used for the adjustment will carry a price index of 100. Now, when the annual costs of another portfolio of the same purchasing category is calculated using the price formula, a mathematical annual spending stemming from the price formula is obtained. If the actual total annual costs of packaging materials are divided by the calculated total annual costs and then multiplied by 100, one obtains the index price for the other portfolio.

Using this kind of indexing will provide exact results on the price levels of different portfolios of the same purchasing category, even when the specifications of these two portfolios are totally different. In addition, each portfolio will receive areas for cost optimisation because spikes are determined at the same time.

Suppliers will often object to this method since external benchmarking is, in fact, now possible thanks to the K-Method. Despite the fact that the supplier and the buyer agree to mutual confidentiality of the agreed prices, there is always the danger that a FMCG buyer will changes jobs and go to another FMCG producer and will use his knowledge of confidential information of price levels. Without the K-Method, a precise benchmarking with the prices of his old employer would not be possible because of the different specifications and volumes of his new employer. With the K-Method, he will only need to safeguard the price formula with its individual prices and apply it to the portfolio of his new employer. If that FMCG company has a higher price level than his previous employer the buyer will be able to put high pressure on his suppliers. However, this objection brought forward by suppliers who disagree with price formulas can be invalidated by the simple fact that the buyer can calculate a price formula himself without the cooperation of the supplier. None of the suppliers, neither the supplier of his old employer, nor the supplier of his new employer can prevent the buyer from making a price level analysis with the K-Method—not even when they refuse to agree to a price formula.

3.3 New Price

3.3.1 Agreeing to a New Price

The new specification of a packaging material, normally in the context of the launch or re-launch of a product, requires in traditional buying practice a new price negotiation with the supplier. In many cases several offers from different suppliers are received and the cheapest chosen.

From this traditional method, it is assumed that there is no relationship between the different packaging materials in terms of pricing. It is a well-known economic fact that the volume determines the price ("economies of scale"). To a certain extent, this means that increased volumes lead to lower prices. When buying packaging materials, it is fair to ask what volume is applied to determine the price. Is it the volume of the individual packaging item for which the supplier is processing a price offer or is it the volume of the entire portfolio of that purchasing group?

It is obviously not the latter. Even if the supplier wins the business by supplying a certain packaging item, it will not justify any claims to deliver other packaging items to the buyer for which he has not (yet) submitted a quotation.

So it seems logical that the supplier only makes an offer for the packaging item requested and will offer a price which makes the production and sale of this packaging item economically viable. Packaging items he already supplies only play a role when these packaging items are very similar to the one for which the buyer has requested a quotation. The buyer will use the existing packaging materials as a price reference to check the quotation for plausibility. If the specifications differ only marginally from an existing packaging item, the price for that packaging item should also differ only marginally. Otherwise, the supplier will be unable to explain the new quotation and may have to reduce his prices, even if he already has made the best offer in comparison to his competitors.

Of course, the supplier will take a general view of his FMCG customer and will evaluate the general situation. A new supplier will also consider potential. This means that the supplier will somehow classify his customer, probably not explicitly but in the back of his mind when he enters his expected contributing margin in his internal offering software. While high-ranking customers receive the best prices, customers with a lower ranking will receive more unfavourable price conditions. The supplier will take into account the size of the customer's business, cooperation and field of business as major points to be considered.

A customer with a large volume in a purchasing category will always get the best prices, regardless of how much of this volume if any is produced by the supplier. The supplier will always keep in mind the potential of this customer.

Also the communication between supplier and customer is an important factor when the supplier defines his margin expectations. If the customer is difficult during negotiations, continuously changes call-offs, even on short notice, lets the supplier down when it comes to disposal of residual stock even if both parties are guilty that these stocks exist in the first place, if he frequently changes suppliers, he will receive a low ranking from the supplier. Such customers will have to pay a premium in terms of high packaging prices for their bad conduct. Sometimes the customer even overdoes his bad habits and the supplier discontinues the supply relationship entirely.

Finally, customers who at first sight seem to be very rich, will pay more than customers who are rated economically limited. The pharmaceutical producer of an expensive patented drug will pay significantly more for a folding carton for this drug than an FMCG producer who requires a similar folding carton in terms of dimensions, cardboard quality and printing technology for a toothpaste tube which he will then deliver to a discounter. This phenomenon of punishing rich customers nd favouring the poor ones seems to be an unwritten rule in supply management d will not be challenged by the K-Method. Only when a buyer from a low cost CG environment moves to a rich life science company bringing the price la and the supplier with him into the new playing field, will the expensive so enjoy the benefits of cheaper packaging. However, it will probably not change the price of the drug, nor will the drug contribute a significantly higher margin just because of a cheaper folding carton.

3.3.2 Process of Defining the Price

From a buyer's point of view, the process of defining a price takes quite some effort and the process must be repeated every time for new or changed packaging materials. Typically, the marketing accounting department is in the driver's seat and collects the different product and material specifications to carry out calculations for each version. Volume estimates for the final product play an important role as annual volumes influence prices for packaging materials and, therefore, also influence the gross margin of the product.

All the scenarios with different products and consequently different packaging material specifications are evaluated first. Every version of a packaging material needs to have a price to carry out a product cost calculation. The purchasing department will provide these prices but also have them reconfirmed by the supplier. Normally, the main supplier for the respective purchasing category will provide prices in the product design phase.

The supplier is confronted with a dilemma since he does not know whether this specification request will be the final one. If it is the final one, he will need to provide a very competitive price in order to win the business against his competitors. If the request is just one of many on the way to the final specifications, he does well to leave some leeway in his quotation to permit the buyer to carry out further negotiations because the buyer will certainly revisit the price once the specifications have become final. Real competitive prices in the design phase of a product are counter-productive for the supplier as these prices define the price level in the market and can potentially spoil the market. No supplier can be interested in such price erosion, especially not when there is no business to win for him during that phase.

Once the buyer has received from the supplier the price for a packaging item variant, he will pass it on to the marketing accountant. He will include the price in the product cost calculation to receive final approval from the marketing department and senior management. In many cases this is an iterative process until the final product version is reached and the final packaging specifications are agreed to by all the parties concerned.

Only when all the final specifications have been agreed to can the buyer negotiate the final prices with the supplier. At this point, the volume requirements are known and the supplier will offer his best price to win the business. Finally, these negotiated final prices are passed on to the marketing accountant to enable him to calculate the final product costs.

This process to determine the final prices is very time consuming, as every specification change and every change of the volume assumption needs reconfirmation from the supplier. The supplier's salesperson will require approval of every

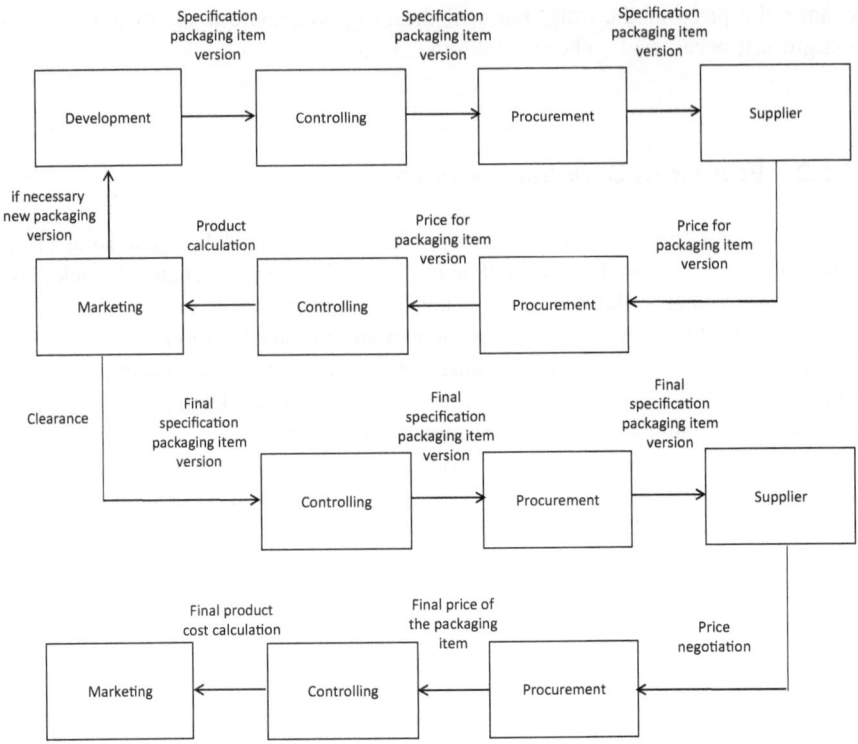

Chart 3.2 Standard process to determine price for packaging materials during launches and re-launches

quotation by his senior management and this will additionally prolong the process (Chart 3.2).

3.3.3 The K-Method for New Prices

With the K-Method there is no need to hold price negotiations whenever the specifications of a packaging material change as a result of a product launch or re-launch. There is a price formula for every packaging category. If there are several suppliers for the same packaging category, the buyer can calculate which of the suppliers offers the lowest price for the new specifications without needing to contact the suppliers or even ask for quotations. Since the price formulas have been agreed to between the buyer and the supplier, there is no need for further consultation. The buyer's tendency to hold negotiations is treated in the Sect. 4.1.2.

It is a major strength of the K-Method that it can calculate these prices which are then passed on 1:1 into the final supply chain. For all the parties concerned, the K-Method is the better solution. Because of the price formula, the supplier is not bothered with providing quotations and the marketing accountant will always work

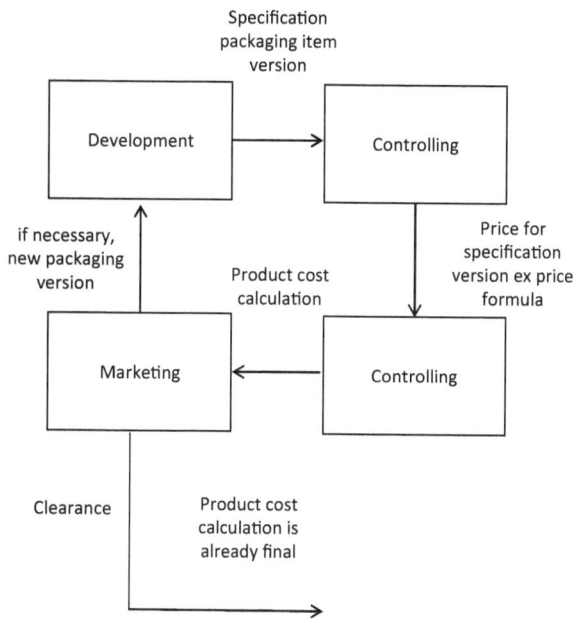

Chart 3.3 Process to determine new prices for packaging materials during launches and re-launches with the K-Method

with the final, originally negotiated prices without having to consult the purchasing department. This is a significant time saver for the marketing accountant.

The following flow chart shows how the process of establishing a price for a packaging item is simplified with the K-Method versus the conventional process (Chart 3.3).

3.4 Tenders

Traditional packaging purchasing uses tenders as a special form of request for quotations ("RFQ"). When preparing a tender, a major part or all of the items in a category are tendered. This means that the present supplier and all challenging suppliers have equal chances to win the business.

Tenders are not popular among suppliers. The present suppliers regard the current and ongoing business as their vested right. For these suppliers, this means that they have something to lose with tenders, either volumes or prices or both. Hence, there is nothing to win for these suppliers. Of course they could also quote higher prices than the current ones which would lead at least mathematically to a higher profit margin. However, it is very unlikely that the existing supplier can maintain the business with increased prices. The buyer will only conduct a tender when he is convinced that prices will drop when he opens the doors to challenging suppliers.

Tenders are not too popular among new, challenging suppliers—although they have nothing to lose. Tenders provide the opportunity for new suppliers to win additional business for themselves but require greater efforts to calculate all the prices required. These efforts are not only to be compared with the number of suppliers invited to the tender—more suppliers reduce the chances of an individual supplier winning the tender. These efforts are also related to the risk that the new supplier will not win the tender in spite of the fact that he might be offering the lowest prices. Instead, the buyer is tempted to use these best prices from a challenging supplier to put pressure on his present supplier to match these prices.

While suppliers react reluctantly towards tenders, they are also not very popular with buyers. Buyers, too, face several problems. First of all the data volume for the tender is massive since they need to collect all the specifications of all the individual packaging items which are the subject of the tender. The supplier will have to decide on a customer-specific price strategy. But he will still have to run all the specifications through his price quotation system which he has set up with his specific margin expectations for the customer. This is actually an idiotic and useless task. Once the buyer has received all the quotations, he will have to enter, supplier by supplier, all the prices into his system typically producing very large Excel worksheets. He will then calculate the total purchasing costs for the entire portfolio and for each supplier individually based on the expected volume and the quoted price of each individual packaging item. In this way, the buyer will determine the winner of the tender who will generally be the one with the most favourable prices. However, the winner will not necessarily provide the lowest price for each packaging item; for some items he may have quoted a higher price than the one quoted by a non-winning supplier.

This observation may lead the buyer to allocate parts of the portfolio to different suppliers, normally those with the lowest prices ("Cherry picking"). In return all the suppliers granted part of the tender portfolio will protest because they will claim that they have quoted prices on the assumption that they would be granted the entire portfolio once they win the tender.

Even when the buyer sticks to the original promise to grant the entire portfolio to the winning supplier, he will face another problem. Since the time the invitation for bids was issued, the portfolio will have probably changed. Some items may have been discontinued; new ones may have been added. Volumes may have changed. The winner of the tender has offered the lowest prices. His weighted average price was lower than the weighted average price of his competitors. This was only possible if he had a competitive advantage over his competitors or, what is more often the case, the supplier was prepared to supply at a lower margin than his competitors. In the latter case, the supplier will have taken a conscious decision and will treat his prices as an investment to increase his market share or in the worst case, to defend his share of the market. In any case, he will try to recover and/or improve his margin. Any specification change made by the FMCG customer will be a welcome opportunity to do so. The buyer chose this supplier to supply the entire portfolio for a certain period of time. This also applies to those packaging items changed after the bid for tender closed. If not only the printing image but also the dimensions and even the printing technology changes, the supplier will take advantage of this opportunity and charge premium prices to bring his entire margin back to normal levels.

This means that with conventional procurement tenders providing for a volume guarantee granted by the buyer, there are no safeguards—nor in particular the possibility of using automatic procedures—to maintain the low prices achieved through the tender, even when the buyer guarantees a time period of only a year. This means that tenders for FMCG packaging materials always run the risk that prices will not be sustainable.

3.4.1 The K-Method for Tenders

It is the target of the K-Method to make tenders for entire portfolios more attractive to all the parties involved. With its price formulas, The K-Method offers the possibility of considerably easier handling for both suppliers and buyers. This reduces work, effort and time for both sides and ensures that prices remain sustainable.

With the K-Method there is no need to communicate the complete portfolio with all the individual item specifications to the supplier. The buyer will only collect the individual prices from the price formula. For the supplier, this represents considerably less work. He will not need to run dozens or even thousands of specifications through his price quotation system. He can concentrate on a handful of key prices which are part of the price formula.

The supplier can review the current prices and might possibly change one or two prices in the price formula.

In many cases, the buyer can also identify the cheapest supplier with one look at the quoted prices of the price formula. The supplier offering the lowest Setup/ Changeover prices, the lowest Run price, and the lowest price for feedstock materials is the cheapest supplier. Typically, the supplier with the lowest fixed order price (Setup, Changeover) will not offer the lowest (Run, feedstock materials). This means from a certain order volume upwards, the supplier with the lowest variable order price will be the cheapest. The buyer can determine the cheapest supplier by calculating the entire portfolio as a scenario or just using a representative portion of the portfolio. With a bit of experience, he can also make a direct evaluation without exploding the prices over parts of the portfolio and making an exact calculation.

In any case, data processing is reduced considerably when the K-Method is applied. Even if the buyer applies the price formula over the entire portfolio, he will not need to enter the individual prices of each packaging item for each supplier. This could add-up to several thousand prices. Instead he only needs to enter the prices of the price formula, a maximum of 40 for each supplier.

In addition, both parties are safe as far as prices are concerned. Launches, re-launches, and volume changes of individual call-offs will not cause problems. For the buyer, the market becomes more transparent which is a great advantage for him but not necessarily a disadvantage for the supplier.

When using the K-Method, it is advisable to link to a tender a volume guarantee for a specific time period, e.g., square metre for corrugated packaging materials. This will also mean that a penalty payment of some kind is agreed to when the volume is not met by the supplier. It may also mean that the buyer will receive a bonus if the volume is significantly exceeded. With such a penalty/bonus agreement the parties emphasize that price arrangements are not made for individual packaging items but for whole portfolios, or at least parts of a portfolio. In the mid-term this will increase the supplier's capacity planning accuracy and he in turn will honour this with even lower prices. This is the only way in which packaging items with relatively small production runs—often priced with unreasonable premiums—can take advantage of the volume of the total portfolio. We call this effect "Internal Leverage".

3.5   Lot Sizes

Obviously, the supplier wants to carry out production runs with large volumes which will save several setups which would be required with small production runs with smaller volumes. This is especially valid when the supplier and buyer have agreed to a price system which does not explicitly specify setups but assumes a standard or minimum production run which the supplier strives to exceed. The fact, that with the K-Method setups are paid for separately by the buyer, does not change the supplier's motivation.

The buyer needs to prevent the supplier from gambling with respect to volumes. This means that the supplier produces more packaging than ordered by the buyer. If at a later stage the identical packaging item is re-ordered by the buyer, the supplier would win on his bet. The supplier could charge the buyer for a setup without actually having carried it out physically because the previous production run covered the volume now required. He would only carry the additional capital and warehousing costs which are normally lower than the setup price.

On the other hand, if the buyer changes the specification of the packaging item for the next production run, the supplier would lose his bet. Now, the supplier would not only lose the capital and warehousing costs but also the entire production costs since the additional packaging items he has produced become residual stock and need to be destroyed since they can no longer be sold.

In the long run, the buyer will have access to more information about the future demand and specification situation than the supplier. If anyone can gamble, it should be the buyer. In the long term, the supplier will recognize the good ordering practices of the buyer and will understand that his chances are not good if he still gambles and produces more than ordered. Hence, only the buyer should gamble which also means that the buyer takes all the chances and risks linked to his assumptions. In the end, the supplier should concentrate on producing the volume ordered and fix his prices in such a way that he has an acceptable margin. Any additional revenue hoped for when very hesitant customers place an order which would permit him to invoice setups that were never actually carried out should not be part of the supplier's equation.

3.5.1   The K-Method for Lot Sizes

The K-Method with its explicit pricing of setups is a precise instrument to reward the customer for his good material disposition. The K-Method is a transparent instrument which permits evaluating the chances of saving setups with higher production runs against the costs and risks of possible obsolete stocks.

With the K-Method the material planner can actually prove that precise project planning for launches, re-launches, and promotions as well as sound forecasting are worthwhile. This means that good planning and acting according to plan are directly measurable financially, while using prices based on mixed calculations always require renegotiation in order to make good planning behaviour profitable for the customer.

It is a target of the K-Method to provide the material planner with greater freedom for his work and especially to free him from the constraints caused by scaled pricing and minimum order quantities. This means that the starting point is the MRP run of the ERP System which provides a list of secondary requirements (materials which need to be purchased from third parties, especially packaging materials—here called requirements). It is now the task of the material planner to generate from this list purchase orders for secondary requirements with the correct prices at optimal cost for packaging materials (which are not necessarily the lowest!).

3.5.2   Reach

One of the main features of the K-Method is to price machine setups separately. This means that when the FMCG customer is aiming at the lowest purchase prices, he will need to cover a long time period with a single call-off. The longer the reach of a packaging item the lower will be its purchase price. Unfortunately, the purchase price is not the only criterion when the results of the purchasing process are being evaluated. Warehousing and capital costs and the risk that the packaging can no longer be used because the specifications have changed or because materials have been damaged during storage so that its further processing becomes uneconomical (e.g., this can be the case with folding cartons), are all factors balancing possible lower purchasing prices.

Normally, materials planning departments are informed of launching and re-launching plans at an early stage so they can plan accordingly. This means they know the first production dates of the products being launched and re-launched and will therefore plan requirements only up to the date of the last production run of the old product. Most FMCG producers change the material numbers of the packaging items when specifications change even when only the

Table 3.2 Example labels—pooling of packaging requirements to a single call-off

Label A, 4 variants
Cost of capital: 8% p.a.   0.154% per week
Cost for warehousing: 10% p.a. 0.192% per week

Week	15 16	17 18 19 20	21 22 23 24	25 26 27 28	29 30 31 32	33 34 35 36	37 38 39	
Requirement in k pcs		400	300	400	300	200	500	
Setup in EUR		765.11	765.11	765.11	765.11	765.11	765.11	
Changeover in EUR		180.00	180.00	180.00	180.00	180.00	180.00	
Run in EUR/k pcs		6.08	6.08	6.08	6.08	6.08	6.08	
Run in EUR		2,433.33	1,825.00	2,433.33	1,825.00	1,216.67	3,041.67	
Total Production in EUR		3,378.45	2,770.11	3,378.45	2,770.11	2,161.78	3,986.78	18,445.69
Merge of two production runs								
Setup in EUR		765.11		765.11		765.11		
Changeover in EUR		180.00		180.00		180.00		
Run in EUR/k pcs		6.08		6.08		6.08		
Run in EUR		4,258.33		4,258.33		4,258.33		
Total Production in EUR		5,203.45		5,203.45		5,203.45		15,610.34
Capital, warehouse EUR		25.27		25.27		42.12		92.65
								15,703.00
Savings in EUR								2,742.69
								14.87%

printing image changes. Changing the material number automatically prevents the MRP from generating too many requirements.

Having said this, only costs for capital and warehousing should be considered when carrying out call-off planning. These costs should be determined by a percentage rate for each of the cost types. Both percentages added together deliver the percentage with which the advanced production should be discounted. In this way, the maximum reach which should not be exceeded can be calculated, as at this point the costs for capital and warehousing exceed the savings in setup and changeover costs. Below an example again based on labels.

In the first line of Table 3.2 the weeks (15 ... 39) of a year are displayed. The line below shows the requirements for a specific packaging item which are calculated through an MRP run. For every production run which takes place every 4 weeks, the supplier will produce and deliver packaging materials meeting the needs of the FMCG customer's production. This means that every 4 weeks, the supplier will also produce the specific packaging item and will charge a setup every time. The total cost for the FMCG customer for the label call-offs for all six production runs amounts to 18,445.69 EUR.

In the lower part of Table 3.2, the same production plan for the finished product is assumed. This time, the production of labels for two production runs of the finished product is pooled. In this way the customer saves every second setup for the production of labels. This means that for six production runs by the customer, only three production runs by the label supplier are required. This results in an additional cost of 92.65 EUR for capital and warehousing, but these additional costs are more than compensated by the elimination of the three setups. Hence, when pooling two production runs into one for label production, the total costs for the customer will drop to 15,703.00 EUR which represent savings of 14.87 % versus having a production run by the supplier for every production run by the customer.

This example illustrates very well that despite assuming high rates for capital costs (8 %) and warehousing (10 %), these additional costs will not be close to the savings which can be achieved when eliminating setups and changeovers by pooling the packaging demands of several customer production runs into a single production run at the supplier. This is true for labels and probably also for many other packaging categories. The savings achieved in our example was 14.9 % and this by pooling only two production runs. When pooling three and more production runs, savings will be even higher. Having said this, it is generally advisable to agree on prices for packaging materials in dependence of volumes. This does not necessarily need to be a price formula from the K-Method, even scale pricing does a reasonable job for this purpose. If such a volume context is ignored when agreeing to prices, the supplier will try, in accordance with the information available to him, to pool as many customer production runs into one of his own production runs. When doing this, costs for capital and warehousing are not his biggest problem. His problem would arise if his customer changes his mind. The supplier always risks having produced obsolete items as a result of the customer's changing the specifications of an individual packaging item. In this case, the supplier will have to destroy these residual packaging materials at his own expense. Because of this risk, the supplier is interested most in collecting information from the customer about his plans for launches and re-launches so that he can manage his own risk better. This information will be useless if the customer changes the supplier in the course of a re-launch. The customer might even not invite the old supplier to make an offer for the launched materials.

To avoid unnecessary discussions as to who must bear the cost for destroying obsolete stocks and as a matter of principle, volume prices should always be agreed to. The best solution is to bind the supplier to a call-off plan where he has no freedom to decide when and how many items he must produce. This decision should be taken by the customer's material planner who is the best informed person concerning future volumes. The supplier should not be invited to gamble on future volumes!

3.5.3 Obsoletes

From what has been said so far it seems that costs for capital, warehousing, setups, and obsoletes are related to each other and are somehow in the same family. Traditional procurement defines all the costs for machine setups which we have priced explicitly as Setup and Changeover as part of the price formula, attributable to the purchasing cost. They are, therefore, the responsibility of Strategic Buying (/ Purchasing). The costs for capital, warehousing, and obsolete stocks is part of Operational Buying which normally is a part of the Central Planning Department. Hence, we have conflicting interests between Operational Buying, also known as Material Planning, and Strategic Buying: Strategic Buying is interested in big production runs by the supplier as this will reduce purchasing costs which is a Key Performance Indicator ("KPI") for Strategic Buying. Operative Buying, being a part of the Planning Department, is interested in very small production runs. Here only low costs for capital and warehousing will be incurred. But far more important is the fact that small production runs by the supplier will reduce the risk of residual stocks which will need to be paid for and destroyed by the customer. The budget for destruction of packaging materials is normally managed by the Planning Department which is measured by using as little of that budget as possible.

This conflict between Strategic Buying and Operational Buying is resolved by the K-Method in the following way: Both departments agree on a formula to calculate the annual volume and the number of production runs by the supplier. Typically, the high volume products are produced every week or every other week; the medium volume products are produced every month and the niche products are produced every second or third month. Products which are produced every half year or just once a year are rare. Once the two departments have agreed on such a classification, they can calculate the standard number of call-offs for each packaging item. These standard number of call-offs are used to calculate the prices of packaging items. When using the K-Method with its price formulas the average call-off volume will be part of the price calculation. This calculation results in actual purchasing prices at planned volumes using planned numbers of setups. At the end of the period, the same calculation is made by exchanging planned volumes by actual volumes. This means that actual purchasing costs are not used to measure Strategic Buying. Instead only actual prices evaluated at planned number of call-offs (using actual volumes) are used, not the actual number of call-offs.

Hence, the planned number of call-offs for setups and changeovers evaluated at the respective prices of the price formula are costs which are used to evaluate the performance of Strategic Buying. At the same time these cost are also used as a budget for Operative Buying. If everything works out according to plan, Operative Buying will affect all of the planned call-offs for setups and changeovers and consequently use the entire budget. At the same time Operative Buying will receive a second budget for write-off and disposal of residual stocks. From experience, you cannot make an omelette without breaking eggs.

The idea of the K-Method is to merge both budgets into a single Setup-CO-Disposal budget: on the one hand, the new budget for setups and changeovers, and on the other the existing budget for packaging material disposal. This has two advantages: whatever the material planning of the Operative Buying looks like, performance measurement of Strategic Buying is not affected. Strategic Buying will be judged on figures which are not influenced by frequent or less frequent call-offs. The second advantage: Operative Buying is now free in its decision process and can evaluate to what extent it should be taking a strategy of less frequent call-offs, saving costs for setups and changeovers, but risking producing obsolete stocks. The chances and risks of this decision are now in a one hand: Operative Buying. The conflict between Operative and Strategic Buying is now resolved.

To perfect this concept, the Setup-CO-Disposal budget can be combined with the capital costs for packaging materials ("Charge on Working Capital", "CWC") which is normally a statistical cost in the P&L which does not have any real invoices reflecting it. This CWC is normally a budget which is allocated by Management Accounting.

3.6   Combination of Packaging Items

Due to the high cost of setting up machines, it is advisable to carry out partial setups during one call-off for packaging materials. To reduce the number of full setups in favour of partial setups, the packaging items need to be produced in a certain sequence to have as many partial setups ("Changeovers") as possible and as a few full setups ("Setups") as possible.

The supplier knows the criteria according to which a call-off of several different packaging items needs to be sorted in order to minimise the number of setups. Traditional material planning will delegate this work to the supplier. This delegation always triggers problems in the subsequent invoice control as it is not clear which material items have been combined with a single setup and which ones have not. This makes a big difference when volume-related prices such as scaled pricing or price formulas according to the K-Method have been agreed to.

The aforementioned mixed calculation problem arises whenever a non-volume related single price for each packaging item has been agreed to. This will cause problems for the supplier if call-off volumes decline. In this case, if he doesn't reduce the production volumes on his side, he will increase the reach of each production run and thus risk producing obsolete stock.

Also, when pricing already contains Setup and Run price elements, a packaging item and its complexity lead to mistakes in invoicing if a full setup is charged while the production of the item in reality was combined with other but similar packaging items.

3.6.1   The K-Method for Combining Packaging Items

Only the combination of prices using Setup, Changeover, and Run permits arriving at a fair total price for packaging items. However, this kind of pricing requires quite a bit of effort in master data management to help invoice control to determine the correct number of setups and changeovers of a call-off. The master data we are talking about is called Format Group which can be explicitly or implicitly defined. This procedure is described in the chapter "Implementation in the ERP-System".

3.7   Dimensions of a Moulds

Nearly all packaging items require specific tools for their production. For labels, folding cartons, and corrugated cartons, these moulds are die-cut moulds, printing plates and clichés. For plastic bottles, jars, and closures, these moulds are extrusion and injection moulding moulds.

Chart 3.4 Injection mould for closure—36 cavities. *Photo*: courtesy of Weener Plastic Packaging Group (WPPG), Weener

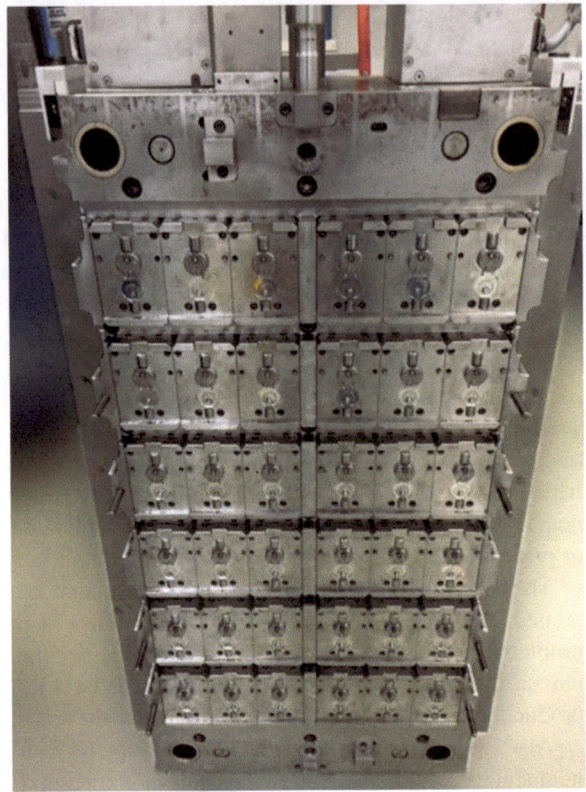

Extrusion and injection moulded moulds, especially, need to be very carefully dimensioned. Big moulds with many cavities are expensive to build and have high maintenance costs, but they have low production costs. On the other hand, moulds with few or just a single cavity are cheaper to build and maintain but involve higher production costs. Basically, two mould scenarios can be sketched as two straight lines using the Y-axis for costs and the X-axis for annual production volumes. The gradient of each line represents the production costs. Any annual volume beyond the crossing point of the two lines will make the mould with the higher number of cavities the more economical one (Chart 3.4).

3.7.1 Applying the K-Method for Dimensioning Tools

Finding the optimal dimensions of a mould is an area where suppliers have a lot of expertise and they normally solve it efficiently. Normally, they compete with other suppliers when a new mould is to be created. Hence, to win the business, suppliers tend to offer moulds at a discount price or even for free. The FMCG buyer can order at competitive market conditions and it is in the interest of all the parties that the moulds have optimal dimensions.

Basically, the K-Method could also be applied to moulds. However, even big FMCG corporations do not normally have a sufficiently large volume of moulds which would permit calculating the price of the moulds based on dimensions and packaging item features using a price formula. The authors admit that in this field, they have reached their limits and need to leave the investigation of prices for moulding unresolved for the time being. We will, therefore, not elaborate further on the ideal number of cavities for extrusion and injection moulding moulds.

3.8 Supply Management Finance

Purchasing Management Accounting/Controlling has two major functions: analysis and forecasting. During analysis, the total purchasing costs within a time period is the main point to be considered. These costs are usually compared with the costs of another period or with the plan for the same period to visualise the differences that result from the variances stemming from volumes, prices and mixed effects.

The Purchasing Department is not normally responsible for purchasing volumes and prefers to be judged by the development of purchasing prices. Thus, volumes and mix variances are combined to show the price variances only. This calculation of prices is normally done by a variance analysis which can then distinguish between feedstock/market variances, conversion variances, and exchange-rate variances.

When looking at packaging items, the question arises how does the volatility of feedstock prices affect the total purchasing cost? The variances can be calculated when the master data of each packaging item also includes the quantity of used feedstock materials.

The exchange-rate variances reflect feedstock materials which are normally denominated in USD or EUR as well as the packaging item in which the feedstock is incorporated. This is calculated normally in the supplier's national currency. When all the prices are taken into account in the original currency and the average exchange rates for a certain period of time is collected, the exchange-rate variances can then be calculated.

The conversion is the part of the price which the supplier charges his customer to convert feedstock materials into a packaging item—i.e., his own production. Eventually, Strategic Buying will want to be measured by the conversion variance as it is the conversion price that he will eventually negotiate with the supplier. The buyer cannot influence the prices for feedstock materials nor can he influence commodity markets. The same applies to exchange rates. Hedging, i.e., the use of financial instruments, may secure a fixed exchange rate for a certain time period and a certain volume. But even hedging will not secure favourable exchange rates in the long run.

The calculation of a variance analysis is made even more difficult by the fact that packaging materials change their material code when specifications change, even when the specifications change only slightly. Technically, the differences of expenditures will then be treated entirely as a volume effect. A price variance can only be calculated when the material code of a packaging item is linked to the material code of its predecessor.

Because of the heavy workload caused by maintaining master data for purchased materials, many FMCG companies refrain from calculating a variance analysis for purchasing. Instead an estimate or extrapolation is made on the basis of supply contracts which due to their low price average volatility (max. +/−4 %), are not accurate enough.

3.8.1 The K-Method for Supply Management Finance

An important parameter when negotiating prices for individual packaging items is the volume expected to be called off over a certain period of time. Traditional FMCG procurement uses 1000 pieces (k pcs) as the Base Unit of Measurement (BuM). For the K-Method this is not sufficient. Alternative BuMs are required which can be calculated automatically from the appropriate master data in the ERP system. For instance, for plastic bottles weight is very important. It is not necessary to differentiate between the different polymer types (HDPE, PE, PP, PET, PVC)— though doing so would lead to even better results when carrying out an analysis. Instead, a standard portfolio bottle can be defined which consists e.g., of 40 % HDPE, 5 % PE, 40 % PET, and 15 % PP. Thus it is not only important to know what the purchased volume in k pcs for a certain time period is, but also what the total purchased net weight is. Using the total net weight and the price for feedstock materials, the price for all feedstock materials for bottles for a determined period can be calculated. When the supplier increases the price for feedstock materials in comparison to a previous period, the price effect resulting from the increase of the feedstock materials price can be calculated directly without requiring a full variance analysis.

With this procedure, every element of the price formula can be evaluated. If the price of that price element changes, the total impact on the respective packaging category can be calculated by using the respective volume. The volume must be in the same BuM as the corresponding price element. This enables the buyer to calculate future purchasing expenditures ("Spend") not only to determine the price effect but also to fully justify the reasons.

Additionally, the buyer himself can make prognoses prior to negotiations. By doing so, he can simplify the calculation process of the annual budget for the entire company. His prognosis can be confirmed later through annual price negotiations with his suppliers. He must take care not to overshoot the prices he has assumed for the annual plan.

Why Should a Company Introduce the K-Method?

4

It should be understood that the K-Method and its price formulas are first of all a theoretical exercise which for outsiders may be a reasonable way to do business. However, insiders such as buyers and especially suppliers are at first resistant to working with the K-Method. This resistance is quite understandable since the K-Method leads to significant time savings for the buyer and especially for the supplier's key accountant. Thus there is a felt threat that introducing the K-Method will require the restructuring of the departments affected. The fact that the K-Method has already been successfully introduced in practice will not reduce such resistance. In this chapter the most frequent points brought forward by buyers and suppliers and how they can be refuted are discussed.

4.1 Buyers' Concerns

4.1.1 General Working Procedures

Even though the authors are understandably enthusiastic about the K-Method because it relieves the buyer from a significant amount of routine work and improves his understanding of the market, we know that this excitement is not fully shared. The buyer is interrupted in his routine and needs to make himself acquainted with a new, rather mathematical technology. Such a change generally generates resistance in the everyday working procedures since the current way of working is always perceived to be the winning approach.

Not only the way of thinking in price formulas is new, negotiating prices using price formulas is also new. Especially the part of negotiations which requires a certain amount of preparation will play a bigger role. This means additional discussions and negotiations and less bargaining with the supplier. Feedstock markets do not only need to be observed carefully, but market trends are now immediately reflected in the prices. Master data plays an even more important role as this data automatically defines prices. An understanding of production

© Springer-Verlag Berlin Heidelberg 2016
D. Kossmann, D. Kossmann, *Complexity Management with the K-Method*,
DOI 10.1007/978-3-662-48244-5_4

technology becomes even more important to show correctly in the price formula e.g., the number of items produced during a production cycle. The expectations of the company in its purchasing department will increase because during value analysis, there will no longer be speculation about the impact on prices when specifications change. The new prices can now be calculated using the price formula and these prices are usually maintained. Even more important, threshold specifications can be deduced which implies that just changing the dimensions of a packaging item by a few millimetres can result in disproportionally high costs or savings. Up to now, the dimensions of packaging items have been constrained only by shelf dimensions and appearance as well as by pallet dimensions. Now, also threshold prices derived from the production hardware of the supplier are taken into account.

Not every buyer will feel comfortable with the new role assigned to him by the K-Method. However, the mitigating factor is that the buyer can now devote part of his time to questions which could not be answered before because of not enough time could be allocated to them. The K-Method will only unfold its full benefits after buyers and suppliers have worked with price formulas for a period of time.

4.1.2 Renegotiations

The K-Method is holistic; it is the only way it can work. One of the strongest features of the K-Method is that new packaging items or specification changes of existing packaging items do not require renewed price negotiations because the price is already covered by the price formula. Especially with new packaging items which are part of a big launch or re-launch, the buyer instinctively feels the need to renegotiate prices. When applying the K-Method he will need to suppress this reaction. In other words, the price formula will deliver a price which the buyer feels can be improved by further negotiations. Perhaps the buyer is not totally wrong since the supplier's salesman is programmed to produce or at least accept a certain level of satisfaction on the side of the buyer by allowing the buyer to bargain the price down.

It needs to be emphasized that this bargaining need by the buyer is already taken into account in the supplier's first quotation. However, the K-Method and its price formula are constructed in such a way that the individual packaging item is not subject to renegotiations. If the buyer ignores this principle, the entire concept of the K-Method is undermined. The price formula is degraded to a price estimation system and all the prices for packaging materials are calculated again on an individual basis. Back to square 1!

Consequently this means that they buyer will not put additional price pressure on the supplier, not even when a big launch or re-launch is planned. This subjective feeling by the buyer of leaving money on the table will provoke resistance by the buyer to the entire K-Method.

But the buyer's discipline will pay off. Once the supplier feels confident that the buyer will not ask for additional price reductions after the price formula has been agreed to, he will be prepared to offer his best price in the price formula. This will then include all the future concessions he was prepared to make during future renegotiations. Having the best price in the price formula is also in the interest of the buyer.

The buyer will argue that the launch or re-launch will change the entire volume base which was originally assumed when agreeing to the price formula. When the volume is reduced the buyer will probably avoid renegotiations as his negotiating position has weakened. However, if the total volume increases he will see the need for renegotiating. This point cannot be refuted easily because every price formula is based on a total volume for the contract period which is typically a year. Should the total volume change because e.g., during a re-launch a product will be offered in a tube instead of in a jar, then the assumption of the total volume of that packaging category is no longer valid. A volume change of a packaging category of +/20 % will indeed require renegotiations. However, this renegotiation should not be conducted for the packaging items which are subject to the launch/re-launch but for the entire packaging category. This means that the prices of the price formula need to be amended. Consequently, the savings stemming from the increased volume are distributed over all the packaging materials of that category and the packaging materials which triggered the renegotiation are not the only ones to benefit from the price reduction. This is fair since all the materials of the category contribute to the total volume of the respective packaging category.

In any case renegotiations of individual packaging items should be avoided, or If need be, the entire price formula must be adjusted. This is the only way that the benefits obtained through the K-Method and its price formula can be maintained.

4.2   Common Concerns of Buyers and Suppliers

4.2.1   Effort and Project Management

First of all, both sides understand that complete packaging portfolios are to be negotiated and not individual packaging items. The negotiations of complete portfolios using a price structure only requires a fraction of the effort of conventional negotiating of all packaging items on an individual basis. It should be kept in mind that every new packaging item—even if only the printing image has changed—needs to go through a process which starts with a request for quotation from a buyer, leads to an offer by a supplier triggering negotiations, and finalises in a new or amended contract in the ERP-System. Up to the point where the price is agreed to, the process can be run in parallel with several suppliers with the intention of eventually agreeing on a contract with the supplier offering the lowest price.

It is understood by buyers and suppliers—as well as by the authors—that only a fraction of the communication between Strategic Buying and the supplier involves price negotiations. However, the time saving is so big that there is a justified fear by buyers and suppliers, but not at management level, that their jobs are endangered if and when the K-Method is introduced. This triggers resistance by those affected which will not be mentioned explicitly. This is also probably the reason why the introduction of the K-Method will fail if the project is run single-handedly. Both buyers and suppliers will bring forward reasons of why in their specific case the K-Method with its price formulas is not the appropriate method to be applied.

In view of this, it makes sense to have an external person manage the project of introducing the K-Method. While the full responsibility for procurement including decisions of supplier selection and prices remains with the buyers, the project manager will need to have the authority to interfere in negotiations and put pressure on the suppliers in order to complete the project successfully. Putting pressure may imply that the business is awarded temporarily to another supplier to underline the customer's commitment to the K-Method. From the author's experience, this kind of pressure is necessary though the K-Method as such does not infringe on the supplier's volume or his margin.

The project manager does not necessarily have to be an external consultant who has experience with the K-Method. An internal person who has already applied the K-Method to one of his packaging categories can be appointed and can now extend his experience to other packaging categories. The authors are also convinced that a buyer who has himself become acquainted with the K-Method can apply it to his category. This buyer will need to have a personality which values innovation in business processes higher than concerns he might have for his own job.

4.3   Concerns of Suppliers

4.3.1   Transparency of Costs

The supplier's biggest concern when introducing the K-Method and price formulas is that he will need to disclose his internal cost calculation. By doing so he is afraid to lose some of his entrepreneurial freedom.

This point brought forward by the supplier is not sustainable for two reasons. First of all, when using such reasoning, the terms "price" and "cost" are mixed up. The real cost of machine hours, labour, depreciation, and overheads are never part of the discussion and will not be negotiated. It is up to the supplier to manage his business and cost most effectively. If he is able to manage these costs better than his competitors, he will be able to convert his cost advantage into extra profit margin for himself without the buyer noticing it or even claiming a portion of it by insisting on lower prices. The K-Method only wants to price specifications, not to simulate the supplier's operations. The supplier's costs, of course, play a significant role when determining prices, but the supplier will adjust prices within his profit margin in accordance with market conditions. Secondly, independent of the fact that we are talking about quoted prices and not the supplier's costs, buyers are usually very well informed about the supplier's cost base. In the early 1970s, in the FMCG industry, big corporations already began to create costs models for the most important categories and to use these as orientation during negotiations. In recent years, so-called "Cost Engineering" has spread widely so that purchasers of smaller

FMCG manufacturers have a good idea of the cost structure of their suppliers. Nothing is being published here that is not already widely known. The K-Method is not causing any damage to the suppliers here.

4.3.2 Inaccuracy

This argument brought forward by the supplier is surprising: On the one hand the suppliers do not want to disclose any information, and on the other, they claim that the price formula is not accurate enough. This is the reason they tend to increase the number of specific production details in the price formula. For example, they claim that when printing labels, the full width of the reel cannot be printed and a margin of 10 mm at both ends are left which will need to be cut off and disposed of at the end of the production. This means that when using an effective printing width of 300 mm, the waste will be 20 mm or 7 %. But what does this production detail have to do with the specifications of labels which is the key to the K-Method and the design of price formulas? Almost nothing. The effective printing width may vary, but not significantly enough to produce a mixed calculation. Instead these margins which cannot be printed should be included in the price for feedstock materials.

When the argument of inaccuracy is brought forward, the question should be asked for which kind of labels is this inaccuracy significant. If a specification feature is not taken into account in the price formula and it has indeed an effect on production costs, the price formula should be extended to consider the different versions of that specification feature. This is a basic principle of the K-Method and not a reason to reject it.

4.3.3 Ability to Combine

The supplier may argue that his quotations are to be seen in the full context of his production portfolio including packaging items for other customers. He suggest that when the specifications of a new packaging item permit the item to be combined in the production process with the packaging item of another customer, he will be able to offer a lower price. This point is often brought forward by the suppliers of corrugated packaging where material waste plays an important role. Here, side runs can be produced which may be used for other production orders. In practice, a product portfolio which can easily be combined does not necessary result in a good production portfolio. The volatility of call-off dates and volumes is so high that it is not worthwhile to store side runs for a long time. Eventually, only those packaging items will be produced which are called-off by the customers. Only such packaging items can be combined which are due for production in the same week. Consequently, to make a price offer, the possibility of combining production runs does not play a role and, therefore, price formulas according to the K-Method can be used without restriction.

4.3.4   Confidentiality

One argument which does not withstand a more accurate examination is the following. If the price formula agreed to between the supplier and the buyer is disclosed to unauthorised persons, this could cause damage to the supplier. Objectively, no damage can be caused. Three different cases are discussed below:

(1) The price formula reaches one of the supplier's competitors. This is indeed an awkward situation because the competitor could take advantage by simply adjusting his own prices accordingly. This means that the competitor will adjust the prices of his price formula in such a way that he will be cheaper than the original supplier and will win the business. The original supplier runs the danger of losing the business he has with his FMCG customer. In defence of the K-Method it must be said that conventional price lists in the hands of competitors have exactly the same effect. When suppliers have access to the prices of their competitors, they will take advantage of this irrespective of how the prices are presented, whether as formula or conventionally.

(2) The price formula ends up in the hands of the supplier's customer B. This, too, is an awkward situation as customer B can calculate his own prices using the terms the supplier granted to customer A. If customer B now discovers that he has been paying more for packaging materials with the supplier than customer A, he will put strong price pressure on the supplier which eventually will lead to the loss of some of the supplier's profit margin. Again, we claim that this is not a specific problem of the K-Method. Generally, no FMCG customer will pass on his prices to his competitor since he also fears the aforementioned effect: Customer A may have a competitive advantage when purchasing packaging materials as compared to his competitor. The last thing customer A wants to do is to call attention to this situation and provide his competitor—customer B—with realistic prices to facilitate customer B's next negotiation round with their common supplier.

(3) The price formula ends up in the hands of an FMCG producer who is currently not a customer of supplier A. Again, the FMCG producer can calculate the prices for his packaging materials using the price formula and compare these prices with the actual prices he currently has agreed to with supplier B who is currently supplying him. If the prices of supplier A are lower than those of supplier B, the FMCG producer may hold negotiations with supplier B hoping to achieve those lower prices with his current supplier since he now has a credible lower alternative offer. He can also switch his business directly to supplier A. In the first case, one of supplier A's competitors will be damaged because this will reduce the competitive advantage he had to date by being able to charge higher prices. In the second case, supplier A will win additional business. Again no damage to supplier A has been caused; quite the contrary. However, we need to quench supplier A's euphoria since as in (2) it is not in the interest of customer A to pass on the price formula to his competitors. But nothing can stop supplier A from sending a price formula to all of his potential

customers in order to win business from them. Conventional tenders for full packaging material portfolios are rare, since they are very labour-intensive for both sides: for buyers as well as for suppliers. Sending out price formulas can be a good acquisition tool for any supplier.

4.3.5 Too Small Portfolio

A supplier's portfolio that has too few different packaging specifications in a single category is considered to be too small. Indeed, when there are more parameters in a price formula than packaging materials in that category, the complexity of this category is very low. In this case the supplier and buyer should agree directly on Setup, Changeover and Run prices without using price formulas. At least the problems linked to scaled pricing are eliminated in such a way.

4.3.6 Dominance of the Supplier

The supplier has a dominant position in the business relationship with the buyer. The buyer more or less buys according to the supplier's price list. This could be the case when the purchasing volume of the buyer is very small, e.g., the investment in a mould of its own for extrusion or injection moulding does not pay off. In this case, the buyer would purchase the required bottles, jars, and closures from a standard mould and negotiate prices individually. Since the buyer cannot change suppliers, the supplier will probably not agree to a price formula. Instead, the supplier will include in his offer the capacity situation of the respective mould which possibly also serves other customers and adjust prices accordingly. In this case, too, price formulas are not sensible.

4.3.7 Advantages for the Supplier

The authors admit that this book was written mostly with buyers in mind. This not only reflects our own biographies but also because the major benefits of the K-Method are with the buyer and not with the supplier. This observation is based more on the experience gathered up to now using the K-Method than on an objective evaluation of the concept and design of the K-Method.

Objectively, the supplier does not lose margin or turnover when the K-Method is applied. It is not inherent to the K-Method to cause prices to fall when it is introduced. If this occurs, it is a result of negotiations between buyer and supplier and is only indirectly related to the K-Method. The assumption by the supplier, that with the K-Method he has to disclose his costs and margins stems from a misunderstanding of the K-Method. It is a fact that with the K-Method neither costs nor margins are disclosed, not even the acquisition costs for feedstock materials which the supplier normally does not buy at index market prices.

Even when the aforementioned is well understood by the supplier, he may remain reluctant to use the K-Method because he still feels constrained in his entrepreneurial freedom. This freedom consists in the ability to provide prices at his discretion every time the buyer requests a quotation for a new or changed packaging item. If a lot of capacity is available he will offer a cut price. If the annual target of the sales representative is already reached, the price will carry a massive margin. The authors doubt that this tactical behaviour towards an FMCG customer is productive in the long run. Instead, in our view, strategic negotiations on annual volumes are the appropriate instrument to develop the supplier's business in the long term.

Once understood by the supplier, price formulas can significantly reduce the workload of his sales department. The supplier only needs to negotiate price levels once a year and can then enjoy a guaranteed utilisation of his production capacity. Negotiations covering periods under one year, with all the surprises inherent in such negotiations, will become history. Not only is a significant reduction of administrative costs visible, but also a significant increase in planning reliability.

If the supplier sees himself at the same level as his FMCG customer and wants to deliver good service, he will not be able to shut his eyes to these points. If he continues to pursue a strategy of obtaining business with dumping prices for high volume items to recover his margin with other packaging items, he will not make friends with the K-Method. However, the authors are convinced that independent of the K-Method, this is a phased-out strategy because the market understanding of buyers has increased over the years.

Having said that, we would like to visualise what the K-Method implies for the supplier (Chart 4.1).

In conventional packaging supply management, the supplier will use a computer programme to make a price offer when quoting for a packaging item. This kind of computer programme is rather complex as the software needs to reflect the

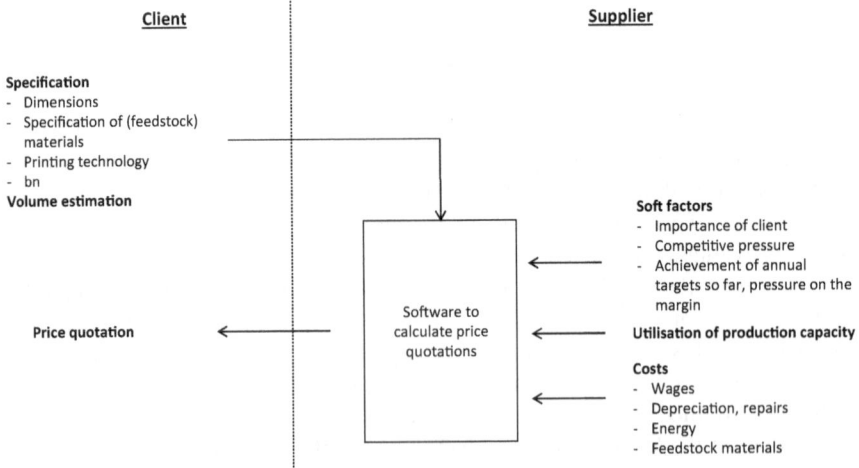

Chart 4.1 Price quotation of the supplier—conventional procurement of packaging materials

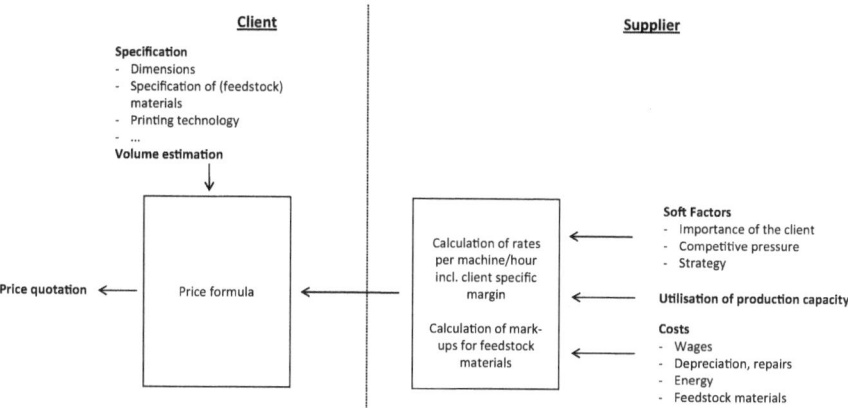

Chart 4.2 Price quotation of the supplier—K-Method

supplier's entire machine park. The software will also make an allocation of the quoted packaging item to machines required in the production process.

The calculation software will have four areas for data input. These data inputs are used to calculate the price. Firstly, there are the specifications of the packaging item which the supplier obtains together with a volume assumption from the FMCG customer. In most cases the volume remains an estimate for orientation and is not binding for the FMCG customer. The production cost parameters are then added to the calculation software. Since the cost parameters are normally calculated only once a year, they need not be entered in the software every time when a quotation is to be made. Thirdly, the latest updated utilisation data is provided. Finally, soft factors which reflect margin expectations are entered.

As mentioned above, the entire process is complex and time-consuming in administrative terms for both sides, the FMCG customer and the supplier. The K-Method now moves this administrative work entirely to the FMCG customer. This move is done by the supplier merging his own costs, utilisation and margins into one price (single price) so as to differentiate this single price from specification features (price formula). The supplier's own costs and margins will always stay merged in a price for a specification feature, thus preventing the FMCG customer from achieving cost transparency. In this way, the FMCG customer receives from the supplier a custom-made price calculation programme which he can operate independently and still not see the supplier's costs and margins. The buyer can now calculate all the prices himself and these prices are negotiated final prices (Chart 4.2).

As long as the K-Method has not become the market standard, the supplier, having made peace with the K-Method, can still take advantage of it. He can provide potential customers with a price formula. The temptation of potential FMCG customers to calculate prices and compare with existing ones will be far bigger than handing out specifications or running a full-size tender. Thus price formulas become an effective sales tool for suppliers to acquire business from new customers.

The K-Method in Other Industries

5

Before starting to explore the potential of the K-Method in other industries outside FMCG and buying categories other than packaging materials, we would like to classify all purchasing materials.

Commodities	This is a category which carries commonly recognized specifications that can be produced basically by anyone who wants to. Because of the high trade volumes, commodities are traded on commodity exchanges. The focus lies on price which is determined by supply and demand.
Specialities/Innovations	This is a category which cannot be produced by other suppliers because its production process or the product itself is protected by patents, copyrights or similar. At least one element needs to be protected while other parts of the production process or product can remain in the public domain. The price is negotiated individually between the supplier and the buyer.
Custom-Made Products	Such products are protected by the customer and can basically be produced by anyone only with the customer's approval. When producing the product the supplier uses commonly available technology.
C-Parts	These are products which are produced according to commonly available specifications or under the protected specification of the suppliers. The customer will be interested in some of the performance features of the product but generally has a certain indifference towards the entire specifications as the purchased products do not play a key role for his own product or for the services he offers.

© Springer-Verlag Berlin Heidelberg 2016
D. Kossmann, D. Kossmann, *Complexity Management with the K-Method*,
DOI 10.1007/978-3-662-48244-5_5

It is obvious that the K-Method is not applicable to commodities since it offers no advantages there. The same goes for C-Parts because the focus of this class lies in minimising effort for negotiation and for the supply process. For C-Parts, a catalogue provided by the supplier is probably the best tool. The supplier will grant the buyer a general discount across all the products of the catalogue and the buyer will manage the supply process with e-procurement tools.

For the class of specialities/innovation, the K-Method will struggle to make a great difference. Theoretically, it is possible to use the K-Method here, too, but the innovation part will have to be negotiated and priced separately. This means e.g., that the buyer of a specific system, say a dashboard for a car producer, will have to pay for the supplier's design/technology through a separate development agreement. This could imply, in an extreme case, that the actual dashboard could also be produced by one of the supplier's competitors.

In this context it will be interesting to take a look at the historic development of industries. Companies that have discovered that the greater part of their value creation lies in innovation, have, over time, outsourced their production, if they have not turned their focus entirely on innovation and become an engineering office. On the other hand, suppliers who saw their core competence and value creation in production will have invested most of their funds in production and will have designed their development departments to act as acquisition tools. This leaves the danger that development work outsourced to suppliers could eventually become very expensive since this development will be paid for by the buyer at premium prices.

Things are different if the supplier of the product can protect himself through a special production technology or perhaps a patent. This is often the case with chemicals. The buyer will have fewer possibilities because an alternative development would probably be too expensive for him.

When materials supplied by the supplier carry a strong element of innovation the supplier will be able to resist the K-Method for the entire material or for remaining price elements which are suitable for the K-Method.

Custom-made products are the true homeland of the K-Method. This is where real complexity exists. When making custom-made products, the same basic technologies are usually applied. For FMCG packaging materials, these basic technologies are: the production of corrugated carton, offset printing, flexo printing, silkscreen printing, extrusion, injection moulding, deep drawing and embossing among others. These basic technologies are applied and combined with different parameters and can be derived from the specifications of the packaging items. Packaging materials for the pharmaceutical industry are produced with the same technologies as for FMCG. Thus, the K-Method can also be used in the pharmaceutical industry in the same way as described in this book.

The K-Method can also be applied wherever basic technologies such as milling, turning and casting are combined, as is often the case in the machine construction industry and in the automotive industry. The same applies to services. Using the K-Method for services is especially interesting for assembly plants. The assembly instructions, take here the role of the custom design which is then translated into the assembly minutes required.

The apparel industry is particularly interesting where sewing patterns are translated into sewing minutes. The buyer can thus concentrate on buying sewing minutes without having to agree to a sewing price for each garment in the collection with sewing companies overseas. The garment industry already works in this a way.

In summary, The K-Method is an agreement between the supplier and the customer on how the specification of a material or a service can be translated into consumption (measured in time, metres, kgs, etc.) by basic technologies so that only prices per time unit must be agreed to for each basic technology. This process can be used for all purchasing categories which have custom-made products or services. Packaging materials for FMCG is just an example.

Outlook on Advanced Technologies

6

Up to now the focus has been on how the K-Method can help automatise the price quotation process for a packaging item by finding a mechanism to translate specification features of the packaging item into single feature prices. The question now arises: Why should the K-Method be only valid for suppliers? Why not for the FMCG customer as well? Can we also apply the K-Method to FMCG producers?

6.1 White Label

It is clear that FMCG products only rarely have cost prices. Instead, they are sold at market prices which reflect the strength of the brand and the general competitive situation of that product category. This results in high margins in the product portfolio of an FMCG producer. However, the situation for white label producers could be different since white label producers have more consistent margins across their portfolio. But the question is still valid for the FMCG producer with branded products, perhaps not for the relationship between the trade and him, but for the internal transfer prices between different units of the same FMCG company.

6.2 Transfer Prices

Transfer prices agreed to between factories and sales units which often are located in other countries can only be defined in a limited way since these transfer prices need to comply with the tax laws of two countries, the country where production is carried out and the country where the final consumer purchases the goods. Only in rare cases is there an agreement on how to deal with unutilised factory capacity. Only in exceptional cases do these transfer prices include machine setup costs as an explicit price item, thus avoiding price volatility when volumes change. Consequently, full-cost pricing on the basis of pieces (or litres or tons) is agreed to. This results in products with high production volumes subsidising those with small

© Springer-Verlag Berlin Heidelberg 2016
D. Kossmann, D. Kossmann, *Complexity Management with the K-Method*,
DOI 10.1007/978-3-662-48244-5_6

production runs because the part of the changeover cost these big volume products carry is bigger than the costs carried by the small volume products. This results in a transfer price which is the same for large volume products and small volume products when packaging and bulk prices per unit are the same. Inversely, complexity costs are underrated because the incremental changeover costs for any additional product variant are spread over all the products of that product group.

Using the K-Method with its focus on volume scenarios and setup price is a good approach to improve transfer price systems by allocating complexity costs in a fairer way.

6.3   Configurators

Once a transfer price system has been built on the basis of the K-Method, exciting new possibilities are opened to Marketing and Sales Accounting. They can plan the supply chain in the same way as the Material Planner manages suppliers. The FMCG supply chain can be treated by sales departments in the same way as packaging suppliers are managed by buyers. Marketing can develop different scenarios of product specifications and will automatically receive a calculation of all the costs including bulk, packaging materials, production costs and logistic costs as well as the resulting profit margin.

If automatic calculation is the target, Product Development will need to provide base formulations of the bulk. These can be adjusted using performance parameters to meet the requirements of Marketing. The required price formulas for packaging materials are provided by the Purchasing Department. The single prices for production, similar to those of packaging materials, are provided by the Production Controller. A product for FMCG can be assembled in the same way as is already being done by the automotive industry.

6.4   Master Data

Such a product configurator will not only offer advantages in time savings and higher calculation accuracy when calculating product costs; it will also open new doors for Master Data Management ("MDM"). While the configurator calculates the different possible versions of a product, the MDM can take the final version from the configurator and feed the master data to the ERP system. Without the knowledge of master data, the configurator will not be able to calculate the costs of the product correctly. The master data of the configurator will need to be enriched by non-cost influencing data such as EAN codes and drawings. Using configurators would not only be a big time saver when it comes to master data but it will also prevent mistakes which often appear in the manual process, e.g., by copying other master data.

The authors are convinced that in the long run, mass producers with a high number of different versions will use configurators for product development, product changes and transfer prices (outside of automotive parts and home appliances, where they already exist). In the near feature those configurators in the form of price formulas will be used only between buyers and suppliers.

Part II

Elaboration

There are no free lunches.
Peter W. Smith

Specific Issues When Designing a Price Formula

7

In the first part of the book the mechanics and the advantages of the K-Method were discussed as well as how to refute the objections of buyers and suppliers. The second part of this book deals with specific issues one come across once a K-Method project has been started and details need to be discussed.

These specific issues stand each one on its own and need to be discussed separately and independently as parts of the price formula.

7.1 Transportation

So far, the costs for the transportation of goods between the supplier and the FMCG customer have not been discussed. In the examples used, transportation was not even mentioned. The prices mentioned, therefore, show the status immediately after production. They are ex works prices (EXW).

In practice, transportation can represent a substantial part of the total purchase cost of packaging materials. Perhaps to a lesser extent with labels, but for plastic bottles, transportation cost can be significant. For plastic bottles, this can play such an important role that it is worthwhile for both parties for the supplier to establish a production site of his own near the production site of the FMCG customer. These so called "through-the-wall" operations are a method in themselves to supply packaging materials. They will not be discussed further here, but through-the-wall operations and the K-Method are equal partners and do not exclude each other.

The transportation of packaging materials is a commodity. It makes no difference which of the parties, the supplier or the buyer, places orders for transportation. Both will pay the same price, neither of them will have a competitive buying advantage. However, it is business practice that the supplier arranges and pays for the transportation and covers the cost over the price of the packaging materials he supplies. One of the reasons for the supplier doing this is that the forwarding agent taking care of the transportation can combine the delivery with the delivery of another of his customers if the delivery does not completely fill a lorry.

© Springer-Verlag Berlin Heidelberg 2016
D. Kossmann, D. Kossmann, *Complexity Management with the K-Method*,
DOI 10.1007/978-3-662-48244-5_7

Since transportation is a commodity, the costs for transportation should be included in the price formula without mark-ups for overhead cost and profit, even when some overheads are involved. Instead, a climate of sharing price information ("Open Book") should be established for transportation. The supplier will ask the buyer only to reimburse the costs he was charged by the forwarder. This means that every packaging item will have to have dimensions which permit calculating the number of packaging items that can be stacked on a pallet. It will be sufficient if this number is not calculated but directly provided by the supplier. The supplier provides the transportation cost per pallet. Transportation cost are variable and depend on the call-off volume, hence they will be part of the run price.

The integration of the transportation costs into the run price is strongly recommended. Otherwise, transportation costs will be charged in a separate line on the invoice. This kind of invoicing should be discouraged because the Management Accounting Department will require effective actual costs for each packaging item. If the call-off has several positions for different packaging items, the transportation costs will have to be allocated by the accountant to each one of the positions. This is a time consuming activity which is prone to causing mistakes. The allocation of transportation costs to packaging materials requires master data of the packaging items which normally is not available or poorly maintained.

7.2   Storage

As promoters of holistic pricing, the authors would prefer not to open a secondary war theatre as such side negotiations would undermine the main negotiations for the price formula. The most frequent secondary issues are payment targets and storage. The latter will be discussed in this chapter.

This chapter deals with custom-made packaging which are packaging items used only by a single FMCG customer. Normally, a custom-made packaging item is defined by a specific printing image or specific dimensions.

In traditional pricing practice, storage is not explicitly mentioned or priced. As a result, some packaging materials are stored at the supplier's location, sometimes for years. Very often, these packaging materials have been produced but not called-off as a result of wrong planning assumptions. In many cases, neither the packaging materials nor the accumulated storage costs have been paid by the customer. The chances that those packaging materials will still find their way into the customer's production are low, and they will probably have to be disposed of. Buyer and supplier are fully aware of the situation when it arises, but the buyer normally will not have a budget for the write-off and disposal of such residual stocks.

Consequently, the supplier will have to pay for capital costs since he has pre-financed the production of these packaging materials and they have not been refunded by the customer. He has also paid for the storage costs these packaging materials have caused. As an experienced supplier, he will have taken such error costs into account. They occur often and are included as part of the overhead costs he has included in his price calculation. Conversely, this means that the customer pays for the errors costs he is responsible for before actually causing the error.

The only way to solve this dilemma is to agree that every production order placed with the supplier is also a delivery order for the same quantity. Generally, the supplier does not keep stocks of custom-made packaging materials and delivers all the packaging materials to the customer at the end of a production run. This is a radical but effective way to keep the supplier free from additional charges. The point brought forward by the customer that he does not have sufficient storage space does not hold when scrutinized because the warehouse of the FMCG customer also holds stock which is outdated and should be disposed of. In this respect radical behaviour which might initially upset the balance sheet and P&L is recommended. It pays off in the long term as it saves costs for all parties.

The supplier must be kept free from all imponderables. This is the only way he can eliminate these from his price calculation which normally include such imponderables to protect his business and his margin. Only then can he offer the best price to the FMCG customer.

7.3 Terms of Payment

It is a widespread custom among buyers to let the supplier down after a production order has been placed by postponing delivery and using him as a warehouse free of charge. Even more widespread is the habit of reopening negotiations on payment terms after prices have been agreed to in order to extend the payment targets. Extending the payment target reduces the working capital of the FMCG customer, enhances his balance sheet and save costs for bank financing which he would otherwise have to bear. Changing payment targets as part of new negotiations under the assumption that the once agreed to prices for packaging materials will not change, is a common practice in purchasing and is perceived by suppliers as bad style.

Both parties are fully aware that tied-up capital comes with a price tag and therefore postponing payment dates come at a cost. However, buyers seem to be reluctant to negotiate payment targets in the context of general price negotiations. The only way for suppliers to defend themselves against extended payment targets after the price negotiations have been completed, is to include this post-negotiation in the price of the packaging materials in the first place. This means that the buyer will pay for extended payment targets even though he may not use it.

In the K-Method a base payment target is used. 14 days could be recommended which leaves sufficient time for a well organised FMCG customer to carry out goods reception control and processing of the invoice for payment. In many countries suppliers grant additional discounts for early payment ("Cash Discount"). Because early payment is often considered to be within 14 days, payment within 30 days would be recommended as a good basis for payments in the K-Method. This leaves the 14 day payment target to incentivise the buyer with an explicit cash discount. In addition, as part of the terms of payment, the supplier would provide an interest rate which the FMCG customer can use to recalculate prices if he wants to extend the payment target of 30 days.

The packaging industry is generally slightly under-capitalised and therefore in many cases the interest rates charged by the supplier will be higher than what the FMCG customer would be offered from a bank to finance his working capital. Most FMCG producers are midsize or large corporations which have accumulated a good capital basis over the years and can therefore receive preferred terms from banks. Seen objectively it is not to the benefit of the supplier nor of the FMCG customer to use the supplier as a bank to finance the FMCG customer over extensive payment targets.

Despite the fact that payment targets can be dynamically calculated and changed by the buyer, buyers and suppliers should agree on a single payment target. Payments made by the customer according to his cash situation should be avoided. This will help the supplier to create a solid and reliable financial plan for his own cash.

By putting a price tag to payment targets the whole matter of payment terms is put on sound and solid feet. Normally, the discussion of payment targets is triggered by the Finance/Controlling Department of the FMCG company and not directly by the buyer. By putting a price on deferred payment targets the FMCG customer will probably eventually opt for the shorter payment target because his previous assumption that payment deferrals are free of charge will not materialise.

7.4   Consignment Stock

Consignment stocks are a very particular case of deferred payment. Such consignment stocks are installed at the FMCG customer's site and are replenished on a continuous basis by the supplier. In order to do so, the supplier receives the FMCG customer's production plans. The supplier will only invoice the customer once the customer has taken materials from the consignment stock to use in his production. The extended payment target is the time between delivery by the supplier and the actual consumption by the FMCG customer's production. Mistakes by the customer's production planners which imply long storage of packaging materials in the consignment stock are all at the expense of the supplier. Eventually, all of these delays and time-offsets will be part of the price quotation which means that the customer will not have an advantage. To the contrary: by using consignment stocks, which is the illusion of always having the required packaging items in sufficient quantities ready at the production line, the buyer loses control over production runs at the supplier's. This is a significant disadvantage as the customer cannot convert improved planning reliability to lower costs of packaging materials. These savings will remain with the supplier.

The additional administrative costs which are caused by consignment stocks are even higher than the loss of control of production at the supplier's. Retrograde consumption calculation, which is standard practice in the FMCG industry, is not sufficient for the invoicing process. Permanent inventory of the consignment stock has to be taken and the corrections made by issuing credit or debit notes. In addition, due to machine setup costs at the supplier's, a strong mixed calculation at the supplier's is the result since pricing according to the K-Method is no longer possible. Even scaled pricing does not work with consignment stock. This will invite the FMCG customer to change his production plans continuously according to his most recent needs. For raw materials, these changes will not cause too many issues because most raw materials are used for a large range of products and are always available in sufficient quantities on stock. But for packaging materials, these changes become a hide-and-seek game for the supplier who will respond with premium prices or walk away altogether from the business because he cannot catch up with his own production planning.

This is the reason consignment stocks for packaging materials in the FMCG industry are not recommend. However, for spare parts for production lines, consignment stocks can make very real sense.

7.5 Delivery Tolerances

In Sects. 7.2 and 7.3, buyers were requested, to negotiate more precisely and especially more fairly, not only in the context of the K-Method. This chapter is now dedicated to the suppliers. It is the practice of many supply agreements that suppliers have a delivery tolerance in terms of volume. Objectively, these delivery tolerances are a nuisance but are hardly perceived as such by the buyer because the delivery volume variances do not influence the purchasing price which is the key measurement for the buyer's performance. If the buyer nevertheless addresses the issue, the supplier will respond with light price increase. The supplier will argue that he, the supplier, is not able to produce the exact call-off quantity due to the technical limits of his production machines. He would then always need to produce a surplus volume in order to avoid underproduction. This would lead to the disposing of goods on his side in order to deliver the exact quantity. This disposal would require him to change the price.

Normally, this kind of discussion does not take place because the material planner—not the buyer—is stuck with the problem of short or excess deliveries. Eventually, the problem is lost in the flood of other problems which the material planner has to face. Packaging material requirements are determined by the MRP run and ordered accordingly. The actual production will stop when the first packaging material is depleted. Normally, packaging materials are planned in such a way that the most expensive packaging material will run out of stock first, which is usually a bottle or a jar. All other packaging materials are calculated more generously usually by bigger loss factors in the Bill of Materials ("BoM"). They are returned to the warehouse at the end of production. (Of course production will also be ended when the bulk materials have been used completely. All the remaining packaging materials are then also returned to storage.)

Because Materials Planning and Production Planning have other priorities, the problem of short or excess packaging deliveries is seldom addressed. In fact, the suppliers use this tolerance to increase their revenue. The excess tolerance is normally always used while the short delivery tolerance is only used in exceptional cases. The disability of suppliers to produce the exact quantity needs to be qualified as a myth. And for the few of them where this is really the case, this should not be charged (with a profit margin!) to the customer but should be carried as a competitive disadvantage by the supplier's margin. Suppliers who manage their own production process well enjoy a nice extra business at the expense of their customer using the excess delivery tolerance. It should be noted that the damage caused to the customer is much higher than the extra profit generated for the supplier. For this reason, the K-Method insists on exact deliveries though it is not required by the methodology of the K-Method. In exceptional cases, the call-off volumes can be rounded off to complete delivery in full carriers, e.g., filling up all the separators of a glass bottle packaging. But allowing a tolerance as a percentage of the entire call-off volume has no economical sound indication because these percentages can be significant when call-off volumes are high. In the worst case, a tolerance can be agreed to on a piece basis as it is a function of poor machine control at the supplier's.

7.6   Optimising the Number of Items During a Production Cycle

For most packaging material types, the number of items produced in a single production cycle is predetermined either by the number of cavities of the tool or through simple algorithms. Based on the dimensions of the packaging item and the restrictions of the production machine, the algorithms calculate the optimal layout and the number of packaging items produced during a production cycle—as was already demonstrated for labels. Because of their simplicity these algorithms can be integrated into the price formula.

One of the big exceptions is folding cartons because they are printed and cut overlapping. This means that the outer dimensions of the silhouette of the flat folding cartons, which is the folding carton before it is glued at the lashes, are not the relevant criteria to determine the number of items per print sheet. The silhouette can be calculated by using the width, the height and the dimensions of the lashes.

The two folding cartons which are shown in Chart 7.1, are placed in such a way that the rectangular silhouette of each carton is used only by the respective folding carton itself. However, it is quite obvious, and usually done in practice, that the two folding cartons can be moved closer to each other to save space and avoid waste. Assuming that the small lashes are not too long, an arrangement of m × n folding cartons on a sheet of cardboard, with "m" being the number of folding cartons in the Y-axis and "n" being the number of folding cartons in the X-axis, can save waste up to (m − 1) × surface of the closing lash. Furthermore, by turning every odd n by 180° it could also be possible to let the gluing lashes overlap.

There are different sizes of cardboard sheets to choose from. Arranging the folding cartons on the sheet is not the only way to optimise printing. Basically, with

Chart 7.1 Press cut folding
carton, source Wikipedia

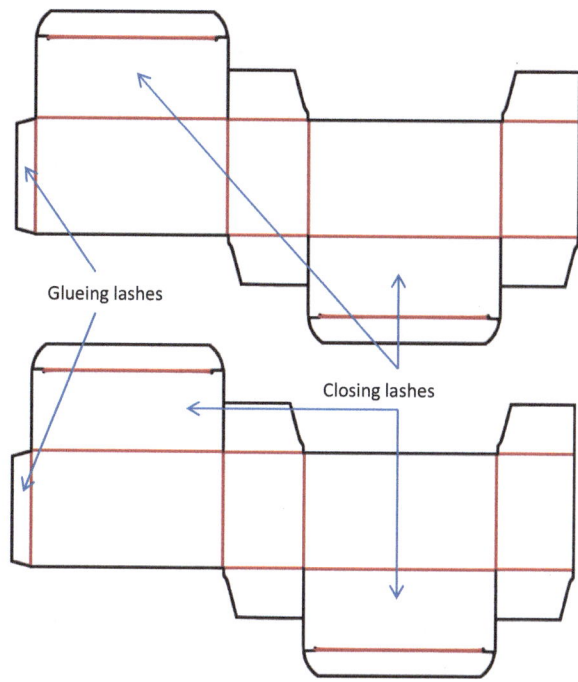

Glueing lashes

Closing lashes

folding boxes, integrating optimisations in a price formula is reasonable even though with a small number of folded boxes the additional waste which would be caused by a silhouette calculation causes no significant impact on the final price.

7.7   Lead Times and "Emergency Call-Offs"

Usually suppliers and buyers agree to lead times which the buyer does not always respect. It is not bad intention on the part of the buyer. He is a victim of his own organisation which continuously changes its production plans in terms of volumes and variants which are to be produced.

As an outsider, this may come as a surprise because FMCG seems to be a predictable business with a constant production triggered by a constant demand. This may be correct but the typical FMCG producer does not supply the market directly but indirectly through the retail trade. Very often, associated companies of the FMCG producer function as agents or links to a local trade partner which means that from the point of view of the FMCG producer, the supply chain to the consumer may have up to three levels. In addition, intensive promotional activities disturb the base line of demand. This is the reason why material planning is always exposed to pressure which it passes on to its suppliers.

Some FMCG companies have agreed on so called "Emergency Call-Offs" with their suppliers, in order to discipline their own organisations. These emergency call-offs are limited in number and are valid for certain time periods or carry an extra charge. They allow the FMCG company to call-off packaging materials at an extra short lead time. When an FMCG customer makes such an emergency call-off, the supplier will reshuffle his own production programme, which often covers several weeks, in order to bring forward the emergency call-off of the FMC customer.

Emergency call-offs can be incorporated into the price formula and the practice of using emergency call-offs can be continued with the K-Method. However, it must be clear to the buyer that using emergency call-offs only reduces the pressure on himself which is nothing more than relieving the symptoms. The actual problem is not solved. Nor does the problem lie with the Production and Material Planning Departments. It is a problem embedded in Sales and Senior Management and they are not prepared to accept economic realities and resolve adequately the triangle of demand, service level and stocks. At least an awareness of the problem by Senior Management is achieved when additional costs for emergency call-offs are agreed to by the buyer and supplier and are approved by Senior Management whenever required.

7.8   Negotiation Cycles

The authors would like to emphasise that price formulas including the respective prices need to stand in a volume context which is guaranteed by the buyer. This implies that a penalty is agreed to between the supplier and the buyer to be paid by the buyer if the call-off volumes do not meet the agreed volume. It is quite reasonable to let the penalty payments start when less than 85 % of the volume is called-off because, normally, the remaining 15 % would not have led to a lower price. The guaranteed volume will increase the planning reliability of the supplier and eventually lead to lower prices.

The prices of a price formula should be subject to negotiations once a year. For this annual negotiation, the new annual planning volume will be provided by the buyer. During negotiations, the buyer and supplier will discuss technological developments in general and the most recent developments and plans at the supplier's production site. Furthermore, assumptions made previously when the price formula was first designed should be reviewed. This goes especially for assumptions concerning waste and machine setups.

If the buyer purchases packaging materials of the same category from several different suppliers using a price formula, he can address the individual prices of the price formula. He will do so especially if one of the supplier's competitors offers special specification requirements (e.g., hot foil stamping) cheaper than the supplier does.

Prices for feedstock materials should be adjusted on a quarterly basis. The easiest way to do so is to link the price to a common available index (e.g., Platts) and add a premium to the price. The simplest way is to use the average price of the previous quarter for the following quarter. This also reflects reality as the supplier normally buys feedstock materials 12 weeks in advance and fixes the prices on the order date of these feedstock materials. Even if this is not done, it should not be a problem as the call-off volumes of the supplier are normally constant over the quarters and time shift effects are compensated in the long run.

So far the focus has been on the principles of the price formula. A basis for discussion with buyers and suppliers who want to fend off the introduction of the K-Method has also been created.

This chapter deals with the implementation of the K-Method. Firstly, the implementation of the K-Method in an ERP system is discussed including the entire business process starting with definition of the price with the price formula, moving on to specifications of the packaging materials, the MRP run, the purchase order and ending with invoice control.

Secondly, selected purchasing categories whose specification criteria can be relevant when designing a price formula are discussed. These reference lists are a good starting point when a buyer and a supplier start a K-Method project and begin creating a price formula. At the same time, these reference lists define the minimum requirements for formal specifications in an ERP system. Once a specification is used in a price formula, this attribute needs to be held in the master data of the ERP system, otherwise the price in the price formula cannot be calculated when generating a purchase order or call-off.

Implementation in the ERP System

8

Most FMCG producers, who have such great complexity in their packaging materials portfolio that makes applying the K-Method worthwhile, usually have SAP as their ERP system. However, the following paragraph is written in such a way that the principles of the K-Method can also be implemented in ERP systems other than SAP.

8.1 The Complete Business Process

After a price formula has been agreed to between the buyer and the supplier, the follow up task is to make all subsequent processes as efficient as possible, if not fully automatic. Savings achieved in the price definition process should not be lost again by labour-intensive business processes for call-offs and invoice control.

The following Chart 8.1 shows the complete business process which is divided into three parts:

1. List of specification features and their possible values, by a category (left).
2. Price formula (above).
3. The process starting from the Material Resource Planning (MRP) run to the purchase order (in the blue box).

8.2 Specification Features

The specifications of every single packaging item needs be stored in the ERP system. Normally, before starting a K-Method project, this is already the case as specifications are used in traditional packaging buying as the basis for price negotiations between the buyer and the supplier. Some FMCG companies store specifications in the form of a text file. For the K-Method this is not sufficient

© Springer-Verlag Berlin Heidelberg 2016
D. Kossmann, D. Kossmann, *Complexity Management with the K-Method*,
DOI 10.1007/978-3-662-48244-5_8

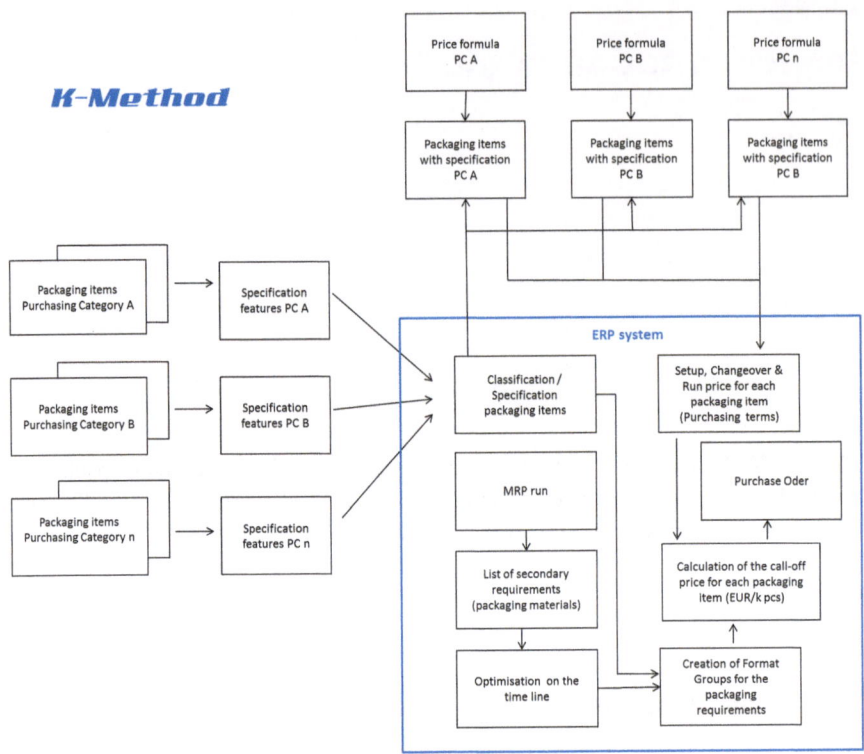

Chart 8.1   Procurement process with the K-Method

because specification features and their respective values are required as an input to determine the price of a packaging item using the price formula.

Before assigning specification features to a packaging item, the packaging item must first be assigned to a category (e.g., labels). This classification will determine the set of features (e.g., height, width, printing technology) to be assigned to the packaging item and which individual values are to be specified. For labels, for instance, these are different specification attributes than for folding cartons. Hence, every packaging item has a certain set of specification attributes in accordance with its classification. In the Chap. 9, some categories are covered with a sample list of specification features which can serve as a reference when designing a price formula.

To facilitate the administration of specification features, meaning the specifications from the point of view of the Packaging Development Department, it is advisable to work with reference specifications; aka master specifications. These reference specifications incorporate all the specification attributes and all the values of a packaging group. Of course, all the packaging items in a packaging group have the same attributes and the same values. Hence, the individual packaging item will not carry the specifications in its master data but will carry a reference to its packaging group (/master specifications). The advantage of using master specifications is a significant reduction in administration work. Many packaging items have the same specifications in principle, but are different from each other only in the printing image. These can share and be part of a packaging group. Again labels are used as an example. From the K-Method's point of view, such labels can be produced for a product with five different fragrance versions and for six different language clusters. This means that there are 30 different labels, all of them with the same specifications. They only differ in their printing image. If any of the specification features changes its value, only one data record needs to be changed—the one with the reference specification—and not the 30 data records of the master data of each label. Once there is a reference link of every label of the reference specification, all the changes of the reference specification will be inherited automatically by every label of that packaging group.

Using SAP, the classification system (module SAP MM) can be used for the definition of specification features. Specification features are administrated even more comfortably in the module "Environment, Health and Safety Management" (SAP EH&S).

8.3   Storage of Prices

It is probably very convenient to store the entire price formula in the ERP-System which would allow dynamic calculation of the exact price of any call-off. However, this will require a significant programming effort. As far as is known, no ERP system is capable of reflecting a K-Method price formula, at least not in such a way that it can be conveniently used on a day-to-day basis. The implementation of a price formula as an Excel worksheet is recommended and this is quickly done.

For every supplier and every category a different Excel Worksheet is required. The price quotations of the supplier are grouped in the first part of the worksheet, the specifications of the packaging items and the expected volume form the second part of the worksheet. The third part of the worksheet will display the prices which consist of three prices: a price for the Setup as an absolute amount, a price for a possible Changeover as an absolute amount, and finally a price for the production run which is quoted per 1000 items (k pcs).

8.3.1   Price Terms

A worksheet as shown in Chart 8.2 three prices for each individual packaging material: Setup, Changeover and Run which can be entered as a price term into the ERP system. Many FMCG customers already work with the "Setup and Run" which means that the change is just an extension of the existing price terms.

Working in such a way, the price formula becomes a preliminary system which permits the automatic calculation of individual prices of all the packaging materials. The quickest way to do so is to prepare a relatively big worksheet

K-Method

Labels ::: Price Formula ::: Sample Prices

Label Specification

Width (Y-Axis)	55	mm
Length (X-Axis)	85	mm
Volumen	100.000	pcs
Print Technology 1	1	1 = offset, 2=Flex, 3 = Silk
Number of colours 1	6	cliches/screens
Number of Change overs 1	4	cliches/screens
Print Technology 2	3	1 = offset, 2=Flex, 3 = Silk
Number of colours 2	2	cliches/screens
Number of Change overs 2	3	cliches/screens
Hot foil blocking?	0	1 = yes, 0 = no
Cold foil blocking?	0	1 = yes, 0 = no
Varnish	0	1 = yes, 0 = no
Polymer type	1	1 = 1362 PE, ...
Ink coverage Technology 1	60	%
Ink coverage Technology 2	60	%

Label Price

Repeat Factor (Nutzen)	6	
Setup Price	1.670,41	€
- of which waste	180,41	€
- of which Change-over	600,92	€
Run on Price	10,48	€ / k pcs
- of which ink	1,21	€ / k pcs
- of which substrate	3,32	€ / k pcs
	0,71	€ / k sqm
Volumen	100.000	pcs
Price EXW	2.718	€
	a	€ / k pcs

Price Components

Currency: EUR

		Offset	Flexo	Silk	
Setup					
Print_Width	Standard print width in order to calculate the repeat factor	350,00	350,00	350,00	mm
Setup	General Machine Setup incl. 3 colours	500,00	300,00	300,00	€
Setup_Waste	Wastage of General Machine Setup incl. 3 colours	150,00	300,00	150,00	m
Setup_Colour	General Machine Setup per addtional colour	60,00	45,00	125,00	€
Setup_Colour_Waste	Wastage of General Machine Setup per additional colour	70,00	120,00	50,00	m
Combi_Setup	Reduction of cost for Set-up when 2 technolgies are used	-30,00			€
Setup_Hotfoil	General Machine Setup for Hot foil blocking	150,00			€
Setup_Coldfoil	General Machine Setup for Cold foil blocking	90,00			€
Change over					
Changeover	Change of one cliche/screen with same colour	60,00	50,00	150,00	€
Changeover_Waste	Wastage of change over per colour	30,00	50,00	70,00	m
Print					
Print	Printing	0,25	0,49	0,90	€/m
Combi_Print	Reduction of cost for Run when 2 technologies are used	-0,08			€/m
Print_HCfoil	Printing of hot or cold foil	0,30	0,20		€/m
Ink					
Ink_price	Price of ink for full coverage = 100%	0,030	0,032	0,400	€ / sqm

Substrates			
1	1362 PE	85 my - white	0,71 € / sqm
2	1428 Poplyefin	88 my - Transparent	0,77 € / sqm
3	1502 PP	60 my - Transparent	0,75 € / sqm
4	1731 PP	60 my - Transparent	0,75 € / sqm
5	1785 PP	60 my - white	0,75 € / sqm
6	1786 PP	60 my - Transparent	0,75 € / sqm
7	Hotfoil		0,40 € / sqm
8	Coldfoil		0,40 € / sqm

Exchange Rates		
1€ =	1,00 EUR	

© i-TV-T AG, Köln 2008

Chart 8.2 Sample price formula for labels as an Excel Spreadsheet

which includes all the packaging items of a category of a single supplier. It will contain all the specifications of each packaging item. Such a worksheet will allow the buyer to calculate all the prices of all the packaging items in one go. The optimal way now is to bulk load the prices calculated in this way into the ERP system.

8.3.2   Standard Prices

When using such a big worksheet, one column should be reserved for the projected average call-off volume. This volume is used to calculate the average call-off price per k pcs which will be used as a standard price for the standard product and which is required for stock evaluation.

8.3.3   Planned Prices

Planned prices are required for budgeting and business planning. Normally, annual business plans are prepared at a moment when the prices for purchased materials for the time period of the business plan are not known because they have not been agreed to yet with the supplier. In such cases the Purchasing Department must make assumptions about the future development of supply markets. These assumptions converted into material prices are called planned prices.

Some FMCG companies have different plans with different planned prices. Especially for packaging materials with its large number of items and variants, this is a complex and time-consuming process. Using the aforementioned method, price formulas can also be used as a planning tool. Assumptions of prices for feedstock materials and transportation as well as about the general market development can easily be entered in the price formulas. In this way, even multi-year plans can be easily managed by the K-Method. The resulting planned prices can be uploaded as Setup, Changeover and Run prices as well as standard prices in the ERP system, preferably again over bulk loading.

In summary, the K-Method is a tool that helps define prices for call-off orders and calculate standard prices for stock evaluation. It also assists in the business planning process providing planned prices for multi-year plans. Using the price formulas will significantly reduce the administrative work involved in preparing such plans.

8.4   Material Planning: Call-offs and Purchase Orders

In the following paragraph the focus will be on call-off prices. These prices are derived directly from the price formula and require a concrete production volume scenario for the supplier.

8.4.1 Material Resource Planning

During an MRP run, the Bill of Materials ("BoM") of the primary requirements (products) are exploded to obtain the secondary requirements, in our case packaging materials. In paragraph "3.5 Lot sizes" the problem of combining material requirements over several production runs was discussed. In the end, it is the decision of Operative Buying (= material planner) to determine the size of the stocks he accepts for a packaging material which, in turn, will define the quantity of the call-off.

8.4.2 Format Group: Setup Versus Changeover

With the help of realistic assumptions of costs for capital and warehousing as well as the risk of the stock becoming obsolete, it is necessary to determine the exact call-off price. By doing so, it becomes apparent that Setup and Changeover conflict with each other. Instead of assigning a setup to every SKU, it is advisable to combine the sequence of SKUs in such a way that instead of generating a full machine setup for each of them, only small changes at the packaging production machine are required so that a changeover only is justified. To automate this arrangement a so-called "Format Group" is required.

A Format Group comprises those packaging items which in an ideal world would be the best produced one after the other at the supplier's. When doing so, they will trigger at the supplier's only one setup and a changeover only for all the following variants. The changeover could consist of the change of a printing plate while keeping printing colours and the die cutter unchanged.

The simplest way would be to define a data object Format Group and assign each packaging material to a Format Group. It would be possible to use a classification attribute or use the Material Group field in the master data (e.g., for SAP). Even if the data is correctly assigned, it does not automatically mean that the price calculation for the purchase order will be correct. The materialisation of Material Groups also requires significant administrative effort to manage master data, and master data is very often already a problem child. Using the K-Method the Format Group will not be explicitly defined and allocated but implicitly calculated using the specifications of the packaging material. This means that two packaging materials PM1 and PM2 belong to the same Format Group whenever specification features PM1/E1 ... PM1/Ei and PM2/E1 ... PM2/Ei have identical values. Of course, the target is to effect the calculations using specification features only. Such specification features are already defined anyway, e.g., folding cartons with the three dimensions and the specifications of the cardboard. Sometimes, an additional attribute needs to be defined, e.g., for plastic bottles. Here the type of mould must be explicitly mentioned.

It is important to agree on the criteria for every price formula with every supplier whenever a complete setup is to be made to determine whether a simple changeover is enough. To make matters even more complicated, buyers and suppliers could end-up defining different types of changeovers, each with a different price.

The dynamic definition of Format Groups requires special software for the ERP system which sorts all the packaging material requirements by supplier, and then sorts all the packaging materials inside these lists, per supplier, in such a way that all the packaging materials of the same Format Group are in sequence.

8.4.3 Purchase Orders

After all the requirement lists for all the suppliers have been created and sorted so that all the packaging materials of the same Format Group are listed in consecutive sequence, the prices for the corresponding call-off volumes need to be calculated. This is done with the following formula:

$n :=$ number of SKUs of the Format Group of a call-off
$\text{Setup}' := \text{Setup} - \text{Changeover}$
$\text{Price PM}j := (\text{Setup}'/n + \text{Changeover} + \text{Run} \times \text{Volume\_PM}j)/\text{Volume\_PM}j$

Alternatively

$\text{Setup}^* := \text{Setup} + \text{Changeover} * (n - 1)$
$\text{Volume}^* := \text{Volume\_PM1} + \text{Volume\_PM2} + \dots + \text{Volume\_PM\_n}$
$\text{Price PM}j := [(\text{Setup}^*/\text{Volume}^*) + \text{Run} * \text{Volume\_PM}j]/\text{Volume\_PM}j$

Every line of the call-off list which carries a packaging item is evaluated and receives a fair share of the Setup cost of its Format Group and its own Changeover cost. The price will be expressed in k pcs. This price per packaging item will be put on the purchase order and should also be invoiced by the supplier with the same amounts.

As there are several different methods to allocate Setup and Changeover costs to the individual packaging items, it is helpful, but not a requisite, to inform the supplier of the method chosen. The supplier can then reconcile the prices on the purchase order and invoice accordingly. But this is not really necessary because the supplier will first calculate the total production price for a Format Group by using the following formula:

$\text{Total\_Price\_FGi} := \text{Setup\_i} + \text{Changeover} \times (n - 1)$
$\quad + \text{Volume\_PM1} * \text{Run\_PM1} + \dots + \text{Volume\_PMn} * \text{Run\_PMn}$

This total price for a format group should be identical with

$\text{Total\_Price\_FGi} := \text{Price\_PM1} * \text{Volume\_PM1} + \dots + \text{Price\_PMn} * \text{Volume\_PMn}$

If this is the case, the supplier can use for his invoice the individual prices (Price_PMj) which are provided by the buyer. Problems will arise only if the supplier produces a different quantity than stated in the purchase order and wants to deliver and invoice this new volume. If the supplier knows the method of allocation, he can adjust the prices and send a corrected invoice. However, in this case, the purchase order price will differ from the invoice price and will require separate approval. This will increase administrative work during invoice control at the FMCG customer's, but he will not need to consult the supplier.

Please note that several ERP systems (e.g., SAP) allow alternative strategies to allocate Setup und Changeover charges to a Format Group (actually the purchase order). The two aforementioned strategies are implemented in the ERP standard of SAP.

8.4.4 Invoice Control

The method mentioned above is designed in such a way that it makes the invoice control easier after the packaging materials have been delivered. The price of the purchase order becomes the relevant price and not the price stored in the ERP frame contract or ERP info record. As a price per packaging item has been calculated for the purchase order a cost allocation does not need to be calculated. The price can be posted directly to the accounts if purchase order volume matches the good reception volume and the invoice price matches the purchase order price.

To avoid mistakes in the master records which can affect the prices in the purchase order and possibly in the invoice if the supplier simply copies them, it is advisable to compare the purchase order price with the standard price. If several production runs are combined or if residual amounts are ordered, the differences between the purchase order prices and standard prices can be considerable so that they are less meaningful as a control figure.

All this means that for invoice control, standard ERP functions can be used and no changes are required in this process.

Specification Features of Selected Categories

9

In this paragraph the specification features of individual, selected categories will be discussed. Such features will probably be part of a price formula since they are cost drivers in the supplier's production. As a reminder: the actual costs are not of interest here. For us, It is sufficient to know that the different values of a specification feature can trigger different costs for the supplier.

9.1 Labels

A certain level of knowledge with labels has been reached because labels have been used as an example to introduce price formulas.

As can be seen, a price formula can be designed even with a short list of specification features. Generally, it is important to keep the list short because every specification feature and its value need to be managed in the master data of the packaging item.

Please note that the list in Table 9.1 only covers standard labels. Holographic labels and leporellos—labels with several layers on top of each other similar to a small book—require additional specification features.

9.2 Plastic Tubes

This example deals with plastic tubes. Laminated tubes which consist of several layers to provide better barrier properties and aluminium tubes have slightly different specification features.

Simple plastic tubes are used mainly for cosmetic products. The decoration of the tube plays an important role. In this example, it is especially hot foil stamping in its different versions which play a role (Table 9.2).

Plastic tubes are produced through extrusion. The extruder melts the polymer granules into liquid plastic and then extrudes the plastic to form a tube. This is a

© Springer-Verlag Berlin Heidelberg 2016
D. Kossmann, D. Kossmann, *Complexity Management with the K-Method*,
DOI 10.1007/978-3-662-48244-5_9

Table 9.1 Price driving specification features—labels

Feature	Data type	
Height	mm	Height of the label
Width	mm	Width of the label
Print_Tec_1	O, F, S	Printing technolgy: Offset, Flexo, Silkscreen
Colours1	n	Number of colours used by the 1st printing technology
Print_Tec_2	O, F, S, -	2nd printing technology (optional)
Colours2	n	Number of colours used by the 2nd printing technology
Hotfoil	yes/no	Hot foil stamping
Coldfoil	yes/no	Cold foil stamping
Varnish	yes/no	Varnish
Substrate	Type	Specification of the substrate used
Ink	p%	Percentage of the surface covered with ink

Table 9.2 Price driving specification features—plastic tubes

Feature	Data type	
Diameter	mm	Diameter of the tube (from a set)
Length	mm	Tube length
HDPE	p%	Portion of HDPE, the remain is LDPE
Colour_T	yes/no	Is the tube coloured or white?
Colour_C	yes/no	Is the closure coloured or white?
Fliptop	yes/no	Closure with a hinge or with screw head
Colours_O	n	Number of colours for offset printing
Colours_S	n	Number of colours for silkscreen printing
Varnish	yes/no	Varnish
Alufoil	yes/no	Tamper evident aluminium foil
Hotfoil_360	yes/no	Hotfoil stamping around the tube
Hotfoil_Logo	yes/no	Hotfoil stamping logo only
Hotfoil_C	yes/no	Hotfoilprinting at the closure

continuous production process. The traditional production technology prints the tube before cutting it. Modern machines cut the tube first and then do the printing. This prevents creating an unprinted fringe at the tube shoulder. This has no influence on the price formula.

The tube diameters which are the supplier's standard tools is what is important here. The most frequent diameters are 30, 35, 40, and 50 mm. In most cases the tube closure will be supplied by another supplier. The most frequent closures used have a hinge and these are a bit more expensive than screwed on caps. When a supplier offers a wide range of closures it is sensible to list all the closures with their individual prices in the same way as is done for the label substrates. The buyer can then include the closure wanted in his specifications.

9.3 Corrugated Outer Cases

Traditional corrugated outer cases are used only for transportation and warehousing purposes and do not appear on the shelf of the retail outlet. These outer cases have a purely logistic purpose. The printed descriptions of the product and the bar codes serve only for information and have no aesthetic ambitions.

In contrast to corrugated outer cases, corrugated shelf cartons, corrugated trays and lids, as well as corrugated displays are used in the service area of retail outlets. These packaging materials have higher printing requirements, which for offset printing require a better paper quality for the paper facing the outside. Alternatively, the outside can be printed separately and then laminated on to the carton. It is not possible to describe these more attractive cartons/trays/displays by their dimensions—height, width, depth—or their corresponding standardised construction numbers (Fefco). Thus the flat carton will have to be designed as a silhouette including waste. The allocation of each individual packaging item on the corrugated base does not normally play a role in price terms compared to folding cartons produced from cardboard (Table 9.3).

Table 9.3 Price driving specification features—corrugated outer cases

Feature	Data type	
Length	mm	Length of the outer case
Width	mm	Width of the outer case
Height	mm	Height of the outer case
Flute_Typ	B,C,D,E ..	Type of flute to calculate the surface
PT_O	KT, TL, ...	Papertype on the outer layer: Kraftliner, Testliner
GR_O	g/m2	Grammage of the outer layer
PT__F	TL, WST, ...	Papertype of the flute
GR_F	g/m2	Grammage of the flute
PT_I	KT, TL, ...	Papertype on the inner layer: Kraftliner, Testliner
GR_I	g/m2	Grammage of the inner layer
Slotter	yes/no	Carton slotter or die cut
Print	S	Print technic, e.g. silkscreen
Colours	n	Number of print colours
Fefco	norm	Type of the case according to Fefco
ECT	n	Edge Crunch Test
BCT	n	Board Crunch Test

The purchasing of corrugated cartons is rather special because two different ways of implementing the K-Method are possible: The purchase of materials and the purchase of performance. When using the material approach, the grammage of the individual layers will be specified by the buyer. The supplier will need to follow these specifications. For this approach, development work will have to be done beforehand either by the supplier or by the FMCG customer's Packaging Development Department to determine the correct grammage. Too low a grammage can lead to the collapse of a pallet when the corrugated outer cases are stacked—the lower cases may not be able to hold the weight. Too high a grammage will lead to unnecessary high costs of the packaging.

Contrary to purchasing according to grammage specifications, the specifications will only define the maximum upset pressure the corrugated outer case should be able to withstand. This pressure can be defined at the edges (ECT) or on the actual board in a pre-processed state (BCT). Some suppliers pride themselves in being able to obtain ECTs and BCTs with lower grammage than their competitors or than can be calculated by standard software. They claim that their secret is in a better glue formulation or in a special selection of papers. In some cases they also base the improvement on a special production process of the corrugated board which allows them to use a lower grammage of the single paper sheets.

Even when suppliers prefer to use performance specifications rather than grammage specifications in the price formulas, using grammage specifications is recommended. When defining the price formula, the supplier's standard grammage and paper combinations must be defined so as not to force the supplier to purchase exotic papers and produce non-standard corrugated boards. Otherwise this would lead to higher production costs and unnecessary setups at his corrugating machine, leading eventually to increased prices.

Part IV

Theoretical Fundamentals

After introducing the K-Method with its price formulas and the resulting business processes, Part IV will explain a few theoretical aspects. These will underpin statements made previously or will indicate directions in which the K-Method can be further developed. The sections of this part, therefore, stand only in a loose connection to each other.

The section "Value Analysis" will demonstrate that technical know-how can be derived from price formulas to specify packaging materials in an optimal way from a Value Analysis point of view. Scaled pricing will again be discussed and its unfitness for purchasing categories with high complexity will be proven. The fact that in cyclic markets prices become inconsistent over time and are, therefore, unfair, will also be shown. The creation of a price formula is described in a more formal way with the perspective of calculating price formulas and prices for any purchasing category.

Value Analysis

Value Analysis focuses on what to buy rather than how or where to buy. The target is to specify a product in such a way that the end customer appreciates the features and is prepared to pay for them. In other words, this means that costly features which do not draw the attention and appreciation of the end customer will have to be eliminated. Or, if the price is to be maintained, how can the specifications of a product be changed in such a way that production costs are reduced without losing the end customer's appreciation of the product?

Some approaches to Value Analysis are based on material standards which influence the life span of the product. The classical question is: if a car's engine lasts 10 years, why should the mirrors last for 20 years? When discussing FMCG products, this approach does not play a big role. Packaging, to the distress of our society, lasts much longer than the product containing it. FMCG products have a best-before date and are normally consumed after days or weeks before expiration. It is obvious, however, that the life span of the packaging should not be synchronized with the actual product, at least not as a primary target. Packaging is supposed to solve logistic, aesthetic and functional problems and should not display signs of deterioration during consumption.

Many technologies to reduce packaging costs are known: changing plastic bottles from HDPE to PET, reducing the wall thickness of a bottle, changing from jars to tubes, downsizing of the lashes of folding cartons, etc. Experienced buyers know them and they will not be discussed further here.

Much more interesting are new approaches of Value Analysis which can be derived—at least by the buyer—from the K-Method. For a Value Analysis project, teams are formed with members of different departments. Typically, these are experts from the Marketing, Packaging Development, Product Development, Production and Purchasing Departments. In many cases a management accountant will work with the team, too. During the project, many alternatives are discussed which need to have a cost evaluation. When the alternative implies a change in packaging, the team will have to make assumptions or request a quotation from a supplier so as to have the confirmation that the envisioned savings can materialise. For this

© Springer-Verlag Berlin Heidelberg 2016
D. Kossmann, D. Kossmann, *Complexity Management with the K-Method*,
DOI 10.1007/978-3-662-48244-5_10

purpose, the K-Method is the perfect tool to calculate different packaging scenarios. The prices determined by the price formulas are already the final prices und no price quotations from suppliers are needed. Not even after a packaging alternative has been approved. The K-Method with its price formulas saves time when carrying out Value Analyses and exact prices are obtained.

Even more interesting are the conclusions in terms of Value Analysis which can be directly drawn from the price formulas. Especially thresholds and the pricing of specification alternatives are obtained sometime in ways which are not obvious.

There are specifications such as the surface of a label which develop parallel to the price. In this example: The bigger the surface of a label, the more substrate the supplier will have to use, and hence, the more expensive the label will be. The relation between the surface (specification) and the required substrate (single price in the price formula) is continuous and linear. This means that if the surface of a label is, say, doubled, the price for the substrate as part of the total label price will also double. Due to the continuity of this relationship, the price as a function of the surface, there will be no exceptions. However, specification features which translate into single prices over a continuous and linear function of a price formula are not of great interest for Value Analysis as they do not deliver any additional insight. Single prices which do not follow continuous functions deliver remarkable thresholds. These are points where the function is not continuous and delivers a local minimum (or maximum). These points require special attention in the specifications as the following example shows.

Label D in our example of Chap. 2 has a specified height of 45 mm. When printing with a maximum print width of 350 mm, seven labels in parallel can be printed effectively using 315 mm of the print width. How does the price of this label change if the specific height is changed? The prices shown in Table 10.1, related to the changed height, are always stated in relation to the originally specified height of 45 mm.

It is worth noting that a continuous increase or decrease of the height of the label does not lead to a proportional increase or decrease of the price of the label. The reason for this is, as already mentioned, the effective printing width. A threshold is reached at 50 mm. A height of 50 mm will permit seven labels to be printed in parallel. A label with a height of 51 mm will only allow a repeat factor of six labels which will lead to a 16.6 % increase in production time taking into account the actual printing only and ignoring the setup time. This will increase the total price of the label by 10.5 %, though the height has been increased by only 4.1 %.

With a printing width of 350 mm, this threshold of 50 mm—and there are other thresholds to consider—can be directly calculated. It is valuable information to know when a label has a suboptimal height in terms of cost. It is worthwhile to investigate whether the label height can be reduced by 2 mm. This would lead to an 8.4 % reduction in the buying price although the height is reduced by only 4.4 %. Perhaps it is only possible to increase the label height. In this case using the full 50 mm would lead to a price increase of only 3.2 % with the height increasing 11.1 %.

Table 10.1 Example labels—prices change when the label height is changed

Label D

Annual volume:	1,750 k pcs
Call-off volume:	292 k pcs

Height	Setup	Run	Call-off		Height
in mm	in EUR/call-off	in EUR/k pcs	in EUR	difference in %	difference in %
57	224.96	5.55	1,842.97	1.1%	3.6%
55	223.91	5.48	1,822.75	1.1%	3.8%
53	222.86	5.42	1,802.54	1.1%	3.9%
51	221.81	5.35	1,782.32	10.5%	4.1%
49	225.05	4.76	1,613.28	1.3%	4.3%
47	223.82	4.69	1,592.89	1.3%	4.4%
45	**222.59**	**4.63**	**1,572.50**		
43	225.13	4.17	1,441.03	-8.4%	-4.4%
41	223.73	4.10	1,420.47	-1.4%	-4.7%
39	222.33	4.04	1,399.91	-1.4%	-4.9%
37	224.17	3.67	1,293.26	-7.6%	-5.1%
35	225.66	3.35	1,204.13	-6.9%	-5.4%
33	223.91	3.29	1,183.21	-1.7%	-5.7%

Of course these thresholds are always to be seen in the context of a favourable pallet scheme. The shelf presentation at the retail outlet also plays an important role. Nevertheless, the thresholds are a helpful and efficient way to determine optimal settings for a Value Analysis. This situation is of particular interest if the price formulas of different suppliers are available. This could mean that the thresholds will differ due to the fact that the suppliers have different production equipment. In this case, more thresholds are available to choose from and a partial portfolio can be allocated to suppliers accordingly.

10.1 Multilingual Labels

This paragraph deals with the popular topic of multilingual labels. Such labels are usually backside labels which are frequently used for cosmetic products. Those labels carry several language blocks placed one after the other, one block for every country. It can be assumed that a consumer will only read the language block of his country and ignore all the others since all the language blocks have the same content.

From the point of view of the packaging costs of multilingual labels, these are not really worth the trouble. This is so obvious that no calculation is required. Assuming a roll of multilingual labels, say with five language blocks: the roll could be cut into five smaller rolls one for each language block. By doing so, five-times as many bottles can now be labelled, the backside labelling costs decrease by 80 %.

In practice, this idea will not meet with too much interest since the number of product SKUs has been inflated to five individual language versions from a single product cluster. Consequently the demands made by one country cannot be met by the product stock of another country. The products are not interchangeable anymore. Within the European Union where country frontiers are not necessarily logistic frontiers, these counter-arguments are not easy to refute.

Scaled Prices

As already mentioned several times, scaled pricing is used to allow different volumes for different call-offs. This is required as the machine setup costs need to be amortised over the production volume. To illustrate, here is an example where the call-off volume is expected to be 75 k pcs. To allow more flexibility to the material planner, additional prices for 50 and 100 k pcs have been agreed to between the buyer and the supplier. The supplier has defined an absolute minimum order quantity of 10 k pcs for which a fairly high price has also been agreed to.

Call-off volume (k pcs)	Price (EUR/k pcs)
10	205.00
50	109.00
75	101.00
100	97.00

Any quantity called off by the customer will have the price of the next lower step price.

The following chart shows graphically the prices which are to be applied depending on the call-off volumes. The Y-axis shows the price per k pcs (Chart 11.1).

The function Volume/Price per k pcs is not continuous, which cannot be expected from scaled pricing either as it contains steps. At the step points, there is point of discontinuity. The problem of scaled pricing is shown better in Chart 11.2 below which shows the volume/price of a call-off.

The points of discontinuity are obvious. More striking is the fact that the absolute cost of a call-off just before a step volume is higher than just after the step point. This means that when an MRP run determines demand for 73.5 k pcs, then in accordance with the price table, a price of 109.00 EUR/k pcs is applicable. This will add up to a call-off price of 8011.50 EUR. On the other hand, if the material planner calls off 75.0 k pcs, he will reach the next step point which would

© Springer-Verlag Berlin Heidelberg 2016
D. Kossmann, D. Kossmann, *Complexity Management with the K-Method*,
DOI 10.1007/978-3-662-48244-5_11

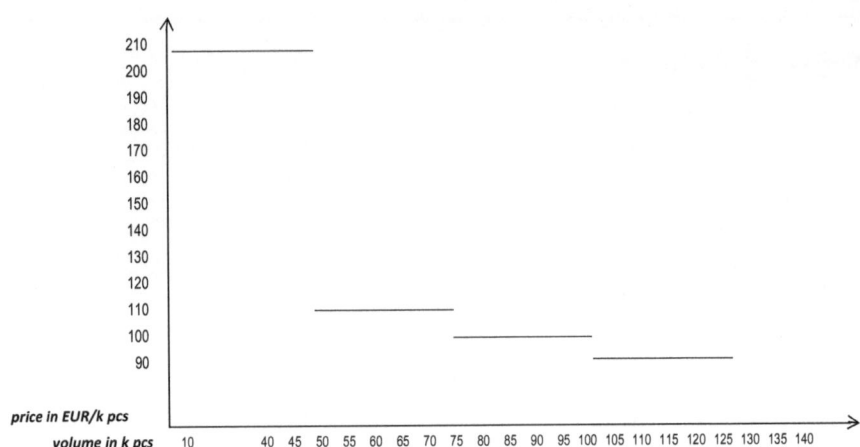

Chart 11.1 Scaled pricing—purchasing prices in EUR/k pcs

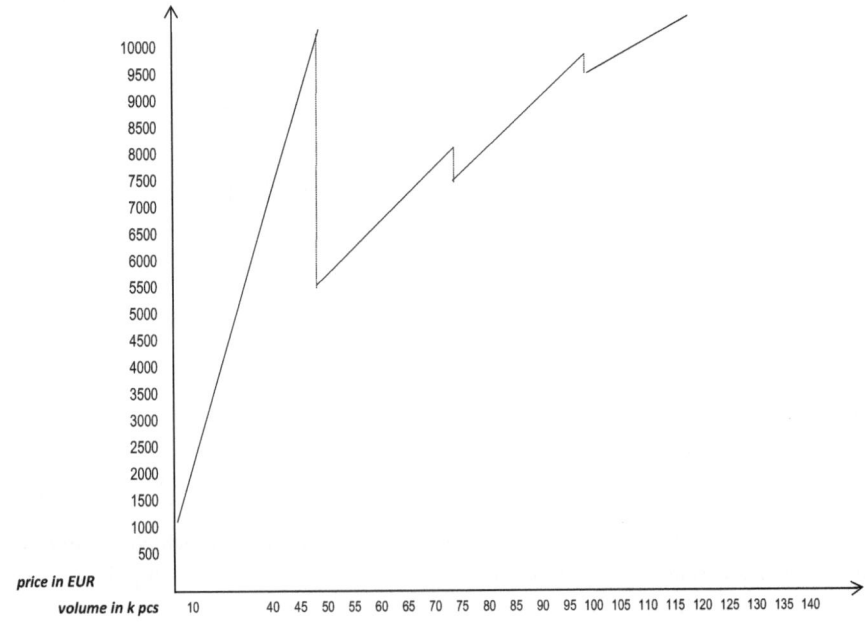

Chart 11.2 Scaled pricing—purchasing prices in EUR

entitle him to a price of 101.00 EUR/k pcs. This would mean a price for the call-off of 7575.00 EUR, with a saving of 5.5 %.

Of course both scenarios are not optimal. On the one hand too high a price is paid, and on the other, a fair price is paid but a quantity of 1.5 k pcs is bought in excess of requirements which either needs to be expensively stored until the next production run or, even worse, will have to be disposed of.

Why machine setups should be paid for separately has already been explained. Thus it is best to do without scaled prices. For the buyer, the conversion from scaled prices to Setup and Run prices is easy to calculate. Two step points are taken and the respective full call-off prices for the volumes required are calculated.

Call-off volume (k pcs)	Price (EUR)
75	7575.00
100	9700.00

Now, a line is drawn between these two points. When this line crosses the Y-axis we obtain the Setup price, the gradient is the Run price:

$$\text{Line\_Definition} \; = \; f\,(X) \; := \; A * X + T$$

This is the equation of a line, with "A" defining the gradient and "T" the crossing point with the Y-axis.

Two points have now been defined with P1 (75; 7575) and P2 (100; 9700) taken from the scaled price list. This leads to a gradient of $(y2 - y1)/(x2 - x1) = (9700 - 7575)/(100 - 75) = 85$. Consequently, the crossing point with the Y-axis is 1200 because $7575\ \text{EUR} - (85\ \text{EUR/k pcs} * 75\ \text{k pcs}) = 1200\ \text{EUR}$.

As shown, the buyer can convert the scaled pricing to a Setup and Run pricing with a Setup price of 1200 EUR and a Run price of 85 EUR/k pcs. The supplier will probably agree to this as his revenue will be exactly the same as before at these step-point volumes. The buyer though, would pay significantly less for volumes which lie between the step points. With an MRP volume of 73.5 k pcs, the buyer would have paid 8011 EUR using the scaled pricing system. Now he will pay only 7448 EUR which represents a saving of 7 %.

Consistent Prices

12

In different parts of this book, consistent prices are stipulated. In the short term, consistent prices have an advantage for the supplier. Inconsistent prices permit the buyer to use spikes in the upper direction. In the midterm, consistent prices are beneficial to both parties as the profit margin of the supplier is spread evenly over all the packaging items he supplies and both parties do not suffer from the disadvantages of combined costing. Consistent prices are prices that are fair at a certain point in time and this is normally achieved by having all the customer's specifications run through the supplier's quotation programme which then calculates a new price. When doing this, all the prices for the packaging items are calculated with the same "soft factors" (see Sects. 4.3 and 4.3.7). This implies that the margin expectations of the supplier will be the same for all the packaging items.

In practice, however, this is not the case. Prices for different packaging items are calculated and negotiated at different moments in time and therefore are also agreed to at different times. From then on, prices for the total portfolio are renegotiated regularly, normally once a year. However, these negotiations do not contain a recalculation of every packaging item. Instead the result of the negotiation is applied to every packaging item of the portfolio.

Chart 12.1 shows the development of the market prices of a purchasing category. The example shows three different packaging items (A, B, and C) which have different specifications but still belong to the same purchasing category. Packaging item A is a bit more expensive than packaging item B which in turn is a bit more expensive than packaging item C. However, the market price developments of these items are the same for all three as all three are part of the same purchasing group.

Because the chosen purchasing category behaves in cycles, different prices at different times are to be expected. As the first quotations for packaging items A, B, and C were calculated at different times, the prices appear at different parts of the price cycle. The quotation for packaging item A, the most expensive, was requested at the peak of the price cycle. Packaging item C, the cheapest, was quoted for the first time at the cycle's lowest point. Hence, in our example, the price differences

© Springer-Verlag Berlin Heidelberg 2016
D. Kossmann, D. Kossmann, *Complexity Management with the K-Method*,
DOI 10.1007/978-3-662-48244-5_12

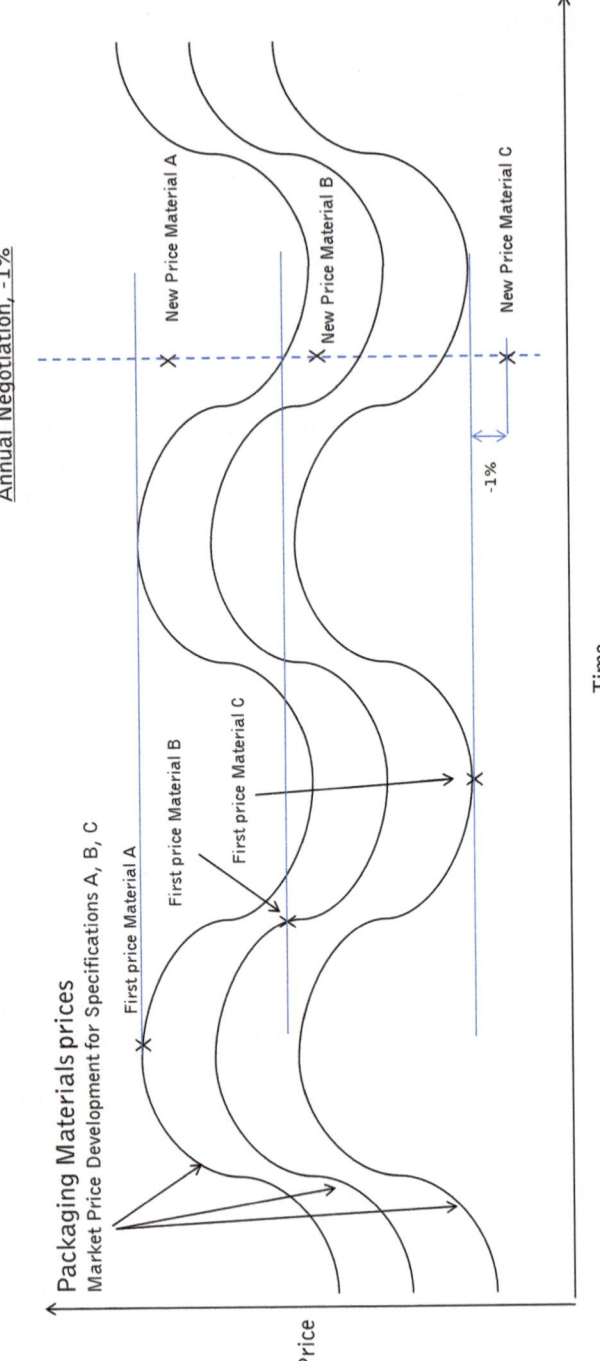

Chart 12.1   Development of the market price in traditional buying

are expected because of the differences in specifications, increased because of the different times in which the quotations were first requested.

When renegotiating prices during annual negotiations, the supplier and the buyer will review the price development of the past year and agree to new prices which will be valid for the following 12 months. In our example, the parties agreed to reduce the prices by 1 % and this will be applied to every single price of every packaging item of the packaging category negotiated.

In Chart 12.1, the blue lines show the old prices for each packaging item. The new price (−1 %) is shown just below. Anyone can see at first glance that the prices are now totally out of line and do not concur with the market price development. Actually, all the prices should be recalculated after the price negotiations. The prices will then again concur with their market price lines. But by applying a general reduction of 1 % to each packaging item, the prices become inconsistent.

Although the market has decreased an average of 1 % for the year, for packaging item A which was quoted at the peak of the cycle, this price reduction is greater. Packaging A will now be purchased at a 1 % lower price, but this decrease is not enough by far. On the other hand, the buyer will also purchase packaging item C at a 1 % lower price although the price for packaging item C should have been increased. The supplier will now be offering item C at a price that is too low. He will only meet his profit margin target if he includes item A into his equation.

In the short term, mixed calculations are the Achilles' heel of the supplier. When the buyer identifies the mixed costing calculations he will withdraw packaging item A from the supplier and ask another supplier to quote for this packaging material. He will surely get a better price than he does from his present supplier. The buyer will generate savings at the cost of his supplier who is stuck with the low margin packaging items he continues to supply.

In the long term, the supplier will include the risk of mixed costing calculation into his prices. This is a risk the buyer has to carry and will lead to higher prices if the buyer remains loyal. But the supplier who pockets this additional profit will not have the planning security of reliable future volumes which he needs for optimum operating and investment planning. Both parties are well advised to refrain from using mixed costing prices. The example given above shows that even when the supplier starts with precise and fair quotations, mixed costing prices creep in through the use of lump-sum price adjustments which could have fatal effects for the supplier in the short term. It is, therefore, better to recalculate the entire portfolio after annual price negotiations as is automatically done with the K-Method.

Derivation of the Price Formula

So far it has been assumed that the supplier and buyer cooperate to develop the price formula. The supplier will then submit a quotation for all the price elements of the price formula. This procedure remains the main principle of the K-Method. This procedure is important and is the only way to apply the K-Method in a sustainable way so that the supplier does not feel thrown off his guard.

For some tasks it might be better to develop a price formula without the supplier. This could be the case for the following:

- Internal benchmarking
- To determine high spikes which leads to renegotiation of prices or change of the supplier
- To determine low spikes to estimate the total saving potential of a purchasing category
- To estimate prices for launches and re-launches.

In such cases, a "normalised price formula" should be derived. A normalised price formula is a price formula where every specification feature has exactly one price element of the price formula as its coefficient which leads to the following format of the price formula.

$$\text{Price (A)} := (M_{A1} * PE_1) + (M_{A2} * PE_2) + \ldots + (M_{An} * PE_n)$$

This means that specification features M_1 to M_n of packaging item A whose price is to be calculated, will only appear once in the price formula and will be multiplied by the respective price element of the price formula. The sum of these products is the price of the packaging item.

Here an example. A totally unknown category is taken which is not related to FMCG packaging materials. Neither the price drivers of this category nor the cost structure are known. There are three materials (Material 1, Material 2, and Material 3) for which a price quotation has been received from the supplier and it is assumed

© Springer-Verlag Berlin Heidelberg 2016
D. Kossmann, D. Kossmann, *Complexity Management with the K-Method*,
DOI 10.1007/978-3-662-48244-5_13

Table 13.1 Example of a fictitious category—specifications and prices

| Material | Feature | | | price |
	A	B	C	EUR/k pcs
Material 1	2	89	no	716.00
Material 2	2	56	yes	1,084.00
Material 3	4	46	no	904.00

Table 13.2 Example of a fictitious category—specifications and prices with converted Boolean values

| Material | Feature | | | price |
	A	B	C	EUR/k pcs
Material 1	2	89	0	716.00
Material 2	2	56	1	1,084.00
Material 3	4	46	0	904.00

that he has provided consistent prices. The supplier has received the specifications of the three materials. Each carries specification features A, B, and C—although each has a different value. Table 13.1, shows the three materials, the specifications and the supplier's prices.

We do not know what the specification features mean. But we can see that different values lead to significant different prices. Before we deduct a price formula we need to convert specification feature C. It seems to be an attribute which shows whether the material has a certain feature or not. The Boolean values "yes" and "no" cannot be transferred to a normalised price formula. Therefore, we must convert the Boolean value "no" to "0" and "yes" to "1".

The table now has the following shape (Table 13.2).

A price formula with the price elements PE_A, PE_B, PE_C must be created to fulfil the following equations.

$$(2 * PE_A) + (89 * PE_B) + (0 * PE_C) = 716,00 \text{ EUR/TST}$$

$$(2 * PE_A) + (56 * PE_B) + (1 * PE_C) = 1084,00 \text{ EUR/TST}$$

$$(4 * PE_A) + (46 * PE_B) + (0 * PE_C) = 904,00 \text{ EUR/TST}$$

This is a linear system of equations with three lines and three unknown elements (PE_A, PE_B, PE_C). The linear system of equations is solved using the Gaussian elimination method (Table 13.3).

The prices of the three price elements of the price formula are obtained as follows (Table 13.4).

Table 13.3 Example of a fictive category—solving of the linear system of equations

Line(Z)	Coefficient (Feature) A	B	C	Result	Calculation	
1	2	89	0	716	Material 1	
2	2	56	1	1084	Material 2	
3	4	46	0	904	Material 3	
4	0	33	-1	-368	Z1-Z2	
5	4	112	2	2168	Z2*2	
6	0	66	2	1264	Z5-Z2	
7	0	66	-2	-736	Z4*2	
8	0	0	-4	-2000	Z6-Z7	
9	0	0	1	500	Z8*-1/4	PEc= 500
10	0	66	1000	1264	Z6 C - use as value	
11	0	66	0	264	-1000	
12	0	1	0	4	Z11 * 1/66	PEB = 4
13	2	356	0	716	Z1 B - use as value	
14	2	0	0	360	-356	
15	1	0	0	180	Z14 * 1/2	PEA= 180

Table 13.4 Example of a fictitious category—price formula

Price formula	EUR/k pcs
PEA	180.00
PEB	4.00
PEc	500.00

A price formula for this category is derived.

$$\text{Price ()} := (A * 180,00) + (B * 4) + (C * 500)$$

Prices for other materials of this category can now be calculated. These prices will define the buyer's expectations when prices are negotiated with the supplier. This does not mean that the supplier will deliver materials at the price calculated, but the probability is very high that he will match the calculated price from the price formula.

13.1   Data Types of the Specification Features

In the example shown in Table 13.1, three different specification features for each material were defined. Two of the specification attributes were numeric values and the third one a Boolean value. In order to convert the specification features and the price to a linear equation system every specification feature must be numeric.

For the Boolean attribute C, the problem was solved by replacing "yes" and "no" by "1" and "0". It works!

More complicated than Boolean attributes are enumeration types. In the first example where price formulas were introduced by using labels as an example, the printing technology appears as an enumeration type. It would be wrong to replace offset printing by "1", flexo printing by "2" and silkscreen printing by "3" for a normalised price formula. There is no linear relationship between the printing technologies. It is highly improbable that flexo printing will prove to be twice as expensive as offset printing, and silkscreen printing even three times as expensive as offset printing. Since a simple substitution does not work here, a different way must be found. The easiest is to split the attribute into as many sub-attributes as the specification features has possible values. Here an example for another new category.

The example in Table 13.5 has a specification feature E which may have the values "gold", "silver" or "chrome". It will not be a good solution to replace these values by 1, 2, and 3. It is better to split attribute E into three sub-attributes, each of them having a Boolean value (Table 13.6).

Table 13.5 Second example of a fictitious category

	Feature							price
Material	A	B	C	D	E	F		EUR/k pcs
Material 1	102	20560	yes	no	gold		2	3,960.00
Material 2	150	930	yes	no	silver		2	913.00
Material 3	50	34000	no	yes	gold		4	4,902.00
Material 4	233	7333	yes	no	silver		2	1,719.30
Material 5	177	18440	no	no	silver		4	2,418.00
Material 6	84	5600	yes	yes	chrome		4	1,335.00
Material 7	55	6400	no	yes	chrome		4	2,418.00
Material 8	120	1000	yes	yes	silver		2	1,335.00

Table 13.6 Second example of fictive category—attributes converted

	Feature								price
Material	A	B	C	D	Eg	Es	Ec	F	EUR/k pcs
Material 1	102	20560	1	0	1	0	0	2	3,960.00
Material 2	150	930	1	0	0	1	0	2	913.00
Material 3	50	34000	0	1	1	0	0	4	4,902.00
Material 4	233	7333	1	0	0	1	0	2	1,719.30
Material 5	177	18440	0	0	0	1	0	4	2,418.00
Material 6	84	5600	1	1	0	0	1	4	1,335.00
Material 7	177	18440	0	0	0	1	0	4	2,418.00
Material 8	84	5600	1	1	0	0	1	4	1,335.00

Table 13.7 Second example of fictive category—price formula

Price formula

	EUR/k pcs
A	2.00
B	0.10
C	400.00
D	2.00
Eg	1,200.00
Es	20.00
Ec	5.00
F	50.00

When the Gaussian elimination method is applied to this new, fictitious category with its prices, the following price elements of the normalised price formula are derived (Table 13.7).

13.2   Linear Functions and Discrete Functions

Another requirement of the normalised price formula is that every specification feature needs to appear once as a coefficient in the price formula. Consequently, the original example for labels will run into trouble with attributes "Length" and "Width". First of all, both are required to calculate the surface of the label. So, a new attribute called "Surface" needs to be defined, which is the product of length and width. The surface is required to calculate the amount of feedstock material required for production. Introducing Surface A, attributes Length and Width could be omitted. But the actual printing is measured in running production metres for which the label width is required. Hence, the attributes Surface and Width will remain. Leaving the attributes Surface and Width is not an obvious choice. Hence, a bit of production know-how is required.

A much bigger problem is the length of the label which is also needed to calculate the number of labels which are to be printed in parallel. For this calculation a discrete function (whole number division) is used. Unfortunately, discrete functions cannot be deducted from normalised price formulas. This means that with discrete functions the deduction of a normalised price formula reaches its limits. However, not using any discrete functions will not mean failure. It will only make the resulting price formula less precise.

13.3   Summary

Deducting a price formula over a linear system of equations is a powerful tool to analyse the existing, conventional price structure of a purchasing category. It can be used to determine the general price level at which the FMCG company is purchasing as well as spike values which indicate opportunities for alternative purchasing and reducing price levels in general.